OCCUPIED

OCCUPIED

An Iraqi translator caught between two worlds shares his story of survival, resilience, and redemption

Waleed & Hannah Hamza

Endorsements

"Waleed has walked the Hero's Journey and lived to tell the tale. Harrowing, funny, heartbreaking, and uplifting, his story will make you feel anger, outrage, sadness, hope, and joy.

Waleed and I served side by side with the Iraqi Special Operations Forces during the bloodiest year of the Iraq conflict. In a place defined by senseless violence and depraved indifference to human misery, Waleed rapidly became a trusted friend. I can say with a high degree of certainty that I am alive because of him. Years later, I was given the opportunity to help him settle in America. As he settled in, America asked Waleed to serve again, and again he obliged. America owes Waleed a debt of gratitude for being a pinnacle example of selfless service at tremendous personal risk.

You will find his story deeply inspiring. It is a story of valor, loyalty, kindness, and grit—bravely staring down cowardice, betrayal, cruelty, and discouragement."

Pete Kofod
Chief Technology Officer, Xiroko

"War is a crucible, a place where the imminent threat of death shapes the lives and destinies of those caught within it. As a US Army Special Forces soldier, I witnessed firsthand the horror and heroism that define such times. Yet, amid the chaos and strife, there are stories that often go untold—stories of extraordinary individuals who, despite the dire circumstances, embody resilience, courage, and an unwavering commitment to a greater good.

This book is one such story. Written by my Iraqi interpreter Waleed, whose bravery and dedication were pivotal during our time working together in Iraq. This memoir offers an unparalleled perspective on the war. His words take us beyond the headlines and into the battlefield, providing an intimate glimpse into the lives of those who endured the war's harshest realities.

As a soldier, I relied on his linguistic skills to communicate and navigate complex cultural landscapes. However, Waleed's role transcended mere translation. He became a trusted friend and fellow brother-in-arms, whom I trusted with my life in a land of danger and uncertainty. Through his narrative, we gain insight into the Iraqi experience—one marked by loss, hope, fear, bravery, and an enduring spirit.

This endorsement cannot capture the entirety of his experiences nor the profound impact he had on our mission and on me personally. But it serves as a testament to the crucial role he and countless others played. Their stories are woven into the fabric of our shared history, and through this book, his voice finds the recognition it so rightfully deserves. Although Waleed never officially earned the title of Green

Beret from the US Army, in the eyes of every Special Forces soldier who worked with him, he is a Green Beret, tested in the fire of combat.

As you delve into these pages, I urge you to reflect on the profound spirit of humanity that perseveres in even the darkest of times. This is not just a story of war; it is a story of friendship, resilience, and redemption.

To Waleed, my brother, thank you for sharing your story. May it inspire and enlighten others, as it did me. I would go to hell and back with you again!"

SFC Wil Ravelo
US Army Special Forces (Ret.)

OCCUPIED
v 1.0

© Copyright 2024 | Waleed and Hannah Hamza
All rights reserved.

No part of this publication may be reproduced, stored in a retrieval system, or transmitted in any form or by any means—for example, electronic, photocopy, recording—without the prior written permission of the publisher. The only exception is brief quotations in printed reviews.

Published in the United States by Sozo Publishing
An imprint of Hamza Creative Group, LLC

Sozopublishing.com

ISBN 979-8-9915676-2-6 (Paperback)
ISBN 979-8-9915676-0-2 (Hardcover)
ISBN 979-8-9915676-2-6 (E-Book)

Edited by Carlene Hill Bryon and Antje Smith
Cover design by Zan Gantt

Disclaimer

*The views expressed in this publication are
those of the author and do not necessarily reflect the official policy
or position of the Department of Defense or the U.S. government.
The public release clearance of this publication by
the Department of Defense does not imply
Department of Defense endorsement or
factual accuracy of the material.*

Dedication

To my wife, Hannah—thank you for walking every step of this journey with me, for enduring the painful process of reliving these memories, and for helping me put this story into words. This book is as much yours as it is mine.

To my children—you are my hope for the future, a constant reminder of why telling this story matters.

To my Mom & my family overseas—you endured so much because of my service—this story is for you, too. As refugees, your strength despite the distance and hardships is a testament to the bond that time and space can never break. I carry the hope that one day we will be reunited.

To all those who have been forced to flee because of war— may this story shed light on your struggles, your courage, and the hope of finding peace once again.

And lastly, to my brothers, both Iraqi and American, who paid the ultimate price—this story is for you, so that your sacrifices will never be forgotten.

Your legacy lives on through these pages.

Author's Note

This story is rooted in times and places where people lived—and still live—with significant risk. To protect these individuals and their families, we've altered some names and details due to the sensitive nature of these experiences. We are profoundly grateful to everyone who contributed to this book. Although painful memories resurfaced for many during the process, sharing these stories was essential to the creation of this book, as they ultimately carry a message of hope.

Wartime is inherently chaotic, confusing, and traumatic. Memories from those times can be as precise as a sniper's aim or as fleeting as a shadow. As we present this story, we ask for your understanding regarding any inaccuracies, which are entirely unintentional. Some parts may seem uncomfortable or simplistic. But please remember: this is not a historical account—it's simply our story as we remember it.

As you explore the story of my life in these pages, know that this journey is not just my memoir—it's a survival guide for others searching for truth, hope, and liberation. My path to faith wasn't sudden or dramatic, like Paul's conversion on the road to Damascus. Instead, it was a long, difficult journey, marked by war, loss, and heartache. But after thirty-two years, I finally stopped fighting and surrendered my life to the one who created me.

This memoir chronicles that journey. It captures the struggles and victories, the moments of despair, and the glimpses of hope. It is a testament to a heart transformed and a life redeemed by the grace of God. As you read, our prayer is that you will encounter the God who created you and discover your original design. Even in the darkest places, God is with you, ready to redeem and restore. He makes everything beautiful in his time—even the messy parts of our lives.

We've also written this book in the hope that it may influence future US policies on crucial issues such as immigration, sanctions, and decisions that shape the future of other nations. Even when a nation enters a conflict, aiming to be a liberator, it rarely understands that it has, in fact, entered that situation as an invader. The problems within and among nations are complex and require a deep understanding for any measure of resolution to occur. There are no easy answers, and the line between right and wrong is often blurred, making the pursuit of justice and peace even more essential.

Waleed & Hannah Hamza

Contents

Foreword

Prologue — 1

PART 1 Chasing Freedom (1982 – 2004) — 7

    Chapter 1: Marked by Saddam — 9

    Chapter 2: Dreaming of a Free Iraq — 25

    Chapter 3: Terp or Traitor? — 41

    Chapter 4: All In — 57

    Chapter 5: Camp Falcon — 65

PART 2 Controlling Destiny (2004 – 2009) — 79

    Chapter 6: On the Run — 81

    Chapter 7: A Mole Among Us — 87

    Chapter 8: Camp Justice — 91

    Chapter 9: Wedding Bliss — 105

    Chapter 10: The Dirty Brigade — 111

    Chapter 11: Navigating the Gray — 127

    Chapter 12: Civil War — Sunni vs. Shia — 137

    Chapter 13: The Surge — 149

    Chapter 14: All is Lost — 163

Chapter 15: I'm the Monster I Feared   175
Chapter 16: Regrets and Remorse   189

PART 3 Rescued by Faith (2010 – 2013)   201
Chapter 17: Another Chance at Freedom   203
Chapter 18: Never Stop Fighting   211
Chapter 19: The Lady in Red   231
Chapter 20: The Man in White   243

PART 4 Occupied by Grace (2014 – 2018)   261
Chapter 21: Love Conquers All   263
Chapter 22: The Great Rescue   279

Epilogue   293
Salvation Prayer   299
Pictures   300
Acknowledgements   307
Glossary of Acronyms   309
About the Authors   311

FOREWARD

Waleed Hamza's book *Occupied* is a VIP backstage pass to the Iraq conflict of the early 2000s. This narrative is told through the eyes of someone who lived and breathed it. This is a unique peek behind the curtain; a rare perspective of events that shaped Iraq, the United States, and the geopolitical landscape.

Having fought alongside Waleed in Iraq in 2006, I can tell you that his intellect, experience, and placement make this memoir a worthwhile and fascinating read. The author's service placed him in grave danger, and he consistently made the decision to serve a cause larger than himself. I have long marveled at the long-term sacrifice that Iraqis like Waleed offered to Iraq, to the United States, and to the cause of freedom. Their commitment should be cherished.

When I returned to Iraq in 2020, I witnessed that the Iraqi Special Operations Forces (ISOF) had endured chaos and the infiltration of their enemies. Today's ISOF trace their lineage to the units in which Waleed and I served. Specifically, the Counter Terrorist Service (CTS), which is now a national treasure, derived much benefit from the groundwork they laid. Waleed and the Iraqi 'terps' made invaluable contributions to ISOF. They not only provided critical interpretation of local language and culture to US forces, but they also battled toe to toe with a formidable enemy.

Waleed's stories of battle, although remarkable, are only one aspect that makes this book an excellent read. The plot, the setting, and the characters are real. But endings like the one in this story seem more suited to a Hollywood script than to real life. Perhaps this is the most magical aspect of this story. I highly recommend this book to anyone interested in military history, leadership, service, and sacrifice, along with everyone who simply loves a great read.

SFC N.R.
19th Special Forces Group (ABN)

PROLOGUE

In the late 1970s and early 1980s, Iraq was a beautiful and safe country. My family lived in Baghdad, the vibrant capital city. Back then, restaurants and cafés lined both sides of the Tigris River, their string lights shimmering on the water, turning the river into a canvas of twinkling colors. Baghdad was enchanting.

You would wake up early in the morning to the aroma of freshly baked bread wafting down the narrow streets. Laughter was a common sound; families gathered and children played at the park without a care. Life was simple, and the community was tightly woven together. It was a place where everyone belonged—Sunni, Shia, Christian, it didn't matter.

In this environment of acceptance and unity, my parents' marriage stood as a testament to love's power to transcend tradition. Unlike most couples of their time, they did not have an arranged marriage. Instead, they met while working at the post office, and their connection was instant. After a few short months, they were married, and their union was founded on love and mutual respect rather than on a familial arrangement.

In September 1980, border tensions between Iraq and Iran sparked a wave of patriotic energy. When the war broke out, there were ads on TV urging people to join the army and defend the country from Iranian aggression. My mother later told me that my father was moved by those ads and decided to join the public army, an auxiliary force formed to support the conventional military.

So, my father took off for training and then to battle. Within a few months—just after I was born—my father was killed in the war. At the time, I was told he was a 'war hero.' But just a few months ago, I learned the truth. Iraq had been the initial aggressor in that war, and my father was just a pawn in Saddam's quest to conquer Iran.

Today, I look back on my experiences with a perspective I didn't have before. When I learned how to plot maps for military actions, I discovered there was a difference between true north and magnetic north. True north is where the North Pole is located, a fixed geographical point. Magnetic north—the north your compass recognizes—is about 1,200 miles away. It matters whether you align your route to true north or compass north.

Over the course of a few feet, it doesn't make a significant difference. Over a few miles, however, it starts to matter. Over many miles and months of travel, it can be the difference between reaching your destination and getting hopelessly lost.

Our lives are often drawn toward a magnetic north shaped by personal history, childhood traumas and the cultures around us. Following that magnetic pull can feel natural and is fine, as long as we're willing to end up far from true north. But reaching true north requires a map and a march that is aligned differently.

Missing true north doesn't require falling prey to an elaborate ruse. More often, it's a subtle drift, one small step off the path, followed by another, and then another. Sometimes, we stray not because we're convinced true north is elsewhere, but because doubt has been sown.

This is what happened to Adam and Eve in the garden of Eden. If the serpent had told them they were wrong about what God had said—that God had actually allowed them to eat from the Tree of Knowledge—it would have set off red flags. But instead, the serpent (said to be more crafty than any other animal) asked a sneaky question: Did God *really* say you couldn't eat from *any* tree in the garden? And the second they engaged with that question, they were drawn into the serpent's universe. The serpent had sown seeds of doubt, and from then on, humanity has been caught in lies propagated by the enemy of our souls over and over again.

God's original design for us was to reflect him in the way we live, loving justice and the truth. That little tiny shift toward deception in Eden led to a huge course change over the history of humanity, from a single lie, to the first murder, to all the wars, lies and sin we live with today.

Recognizing the truth among all the deceptions is not easy, especially when living under a dictator. So much of my perspective was shaped by propaganda until I oriented myself to true north, discovering the God who created true north, who created everything in this world, who created you and me. And since then I have learned that, as long as we stay oriented to Him, we will make it to our destination without getting off course.

But when this story begins, I didn't yet know true north. So instead, magnetic north pulled me off course time and time again.

Prologue | 5

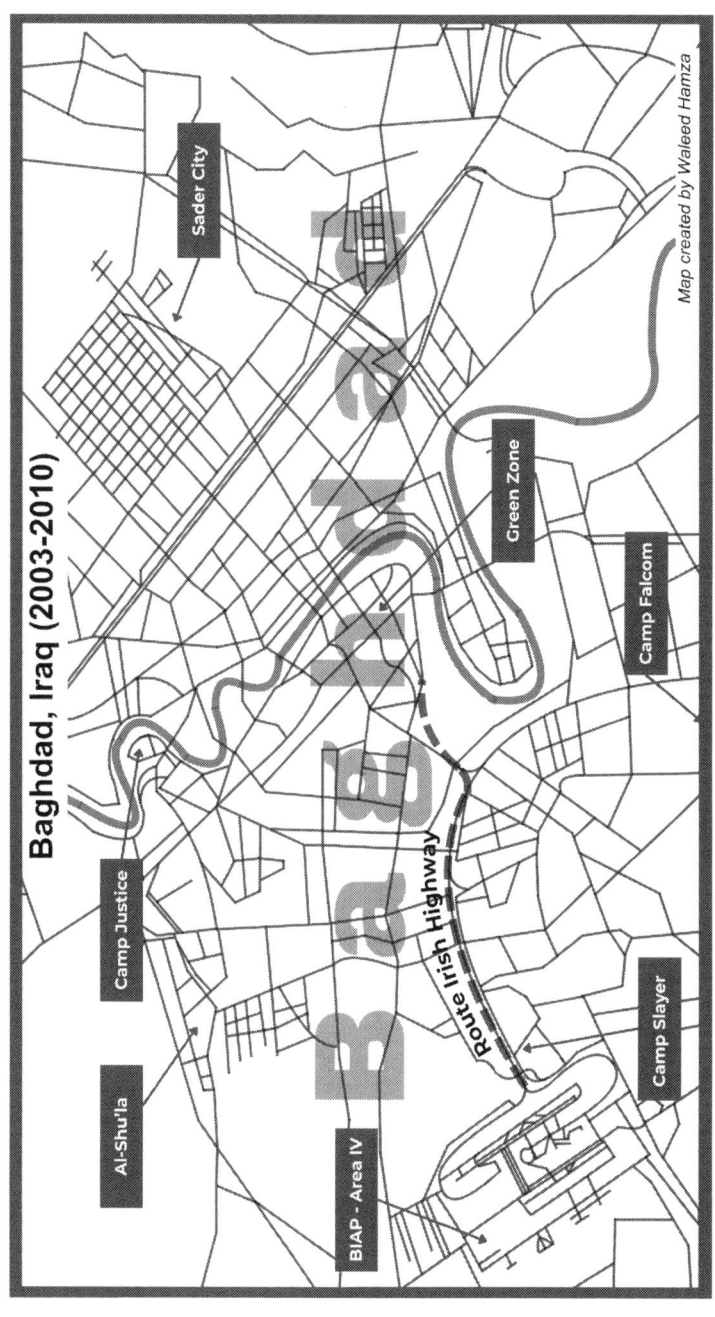

PART 1

Chasing Freedom

(1982 – 2004)

1
MARKED BY SADDAM

My life was marked from the very beginning. First, by the God of the Universe who marked me long before I knew him. Next, by Saddam Hussein. And lastly, by my allegiance to the US military. I lived as a marked man with a bounty on my head and darkness growing in my soul.

This story begins in 1982. My maternal grandfather, a diplomat, had been assigned to work at the Iraqi Embassy in Cairo, Egypt. His nineteen-year-old daughter, my mother, had recently been widowed when the love of her life, my father, was killed in the Iraq-Iran War. The plan was for my grandfather to take my mother and me, a fourteen-month-old baby, with him to Cairo.

Just as we were boarding the plane, soldiers in dark olive green uniforms stopped us, addressing my grandfather.

"The president wants to speak with you on the phone," they said.

So my grandfather and my mother, with me in her arms, followed the soldiers back into the terminal where one of them opened a direct phone line to the palace and handed the phone to my grandfather.

"Mr. President, what can I do for you?" he asked.

President Saddam Hussein was indeed on the other end of the line.

"I have the family of the war hero Khalid here with me," Saddam explained. "And they have asked me to bring their grandson back to them. As parents who just lost their son, I think it is their right to have their only grandson with them. And his mother should stay as well."

My grandfather, trained diplomat that he was, said the only thing he could say.

"Mr. President," he replied. "I believe that you will take care of my daughter, and I trust you in that matter. I'm going to leave her in your care."

And so, my mother and I were taken to the palace to meet Saddam.

Saddam's Republican Palace in Baghdad stood in stark contrast to the rest of the city. It was a stunning example of opulence and grandeur. The architecture was impressive, with a luxurious interior that boasted lavish decorations. Marble floors gleamed underfoot, and granite walls stood solid and imposing, adorned with intricate paintings. The palace featured swimming pools and man-made lakes, carefully stocked with special breeds of fish. The surrounding greenery was meticulously maintained, with well-planned gardens that showcased tall palm trees lining the streets. Many of these trees bore fruit, adding to the lush, almost surreal beauty of the palace grounds.

At the palace, we were escorted to a waiting room, the air thick with uncertainty. Then, after a while, we were called in. My mother must have wondered what awaited her behind those doors—was more upheaval looming on the horizon?

"Samara, the president is ready for you."

My mother clutched me in her arms as she hesitantly walked towards the president. Saddam was sitting at the front of the room. My paternal grandparents, four of my uncles, and my aunt sat off to the side.

Saddam spoke.

"Welcome, wife of our war hero."

She stood waiting, the tension of the moment grating on her resolve to stay collected. The room was silent and heavy as everyone waited for Saddam's next move.

Suddenly, my mother began to sob, the grief cascading over her like a landslide. This was too much, and suddenly her sobs erupted into screaming.

"You've already taken away my husband and killed him," she shouted. "And now you want to take my baby and destroy my life!"

My mother's reaction was understandable given she was a grief-stricken teenage mother who had lost her husband only weeks earlier. She was also desperate to move out of the family home of her late husband because her in-laws treated her harshly. She was seen as a housekeeper not a family member. Egypt was her only chance of escape.

Before she finished saying all that was on her mind, Saddam's personal aide, Abid Hamid Mahmud, grabbed her by the shoulder and squeezed so hard he almost broke her collarbone.

"Let her go," Saddam barked. "She is the wife of a hero, and we don't treat war hero's wives like that."

He looked at the guard.

"Bring Waleed to me."

The guard took me from my mom and handed me to Saddam. The president sat me on his lap.

"Bring some toys for the boy," he commanded.

Saddam again looked at my mother.

"They should have access to Waleed, their grandson, given that they've just lost their son, too. We have already discussed a resolution for the situation."

Saddam nodded towards my uncles. Suddenly, all four of them lined up in front of Saddam, as if they were in the military and he was their general. Saddam pointed to one of them, the eldest of the brothers, and then quickly looked back at my mother.

"You will marry him," he declared. "He is a lawyer and is well established financially. It's settled."

My mother was in shock, but she didn't dare oppose the dictator of her nation. So right there on the spot, she was forced to marry my uncle to keep me in the family. Saddam then sent my mother away with a large sum of money as compensation for her late husband's death and as a gift for me. But no amount of money could fix the mess she was now in as a grieving widow in a loveless marriage.

The first thing my mother did with the cash gift from the government was buy a house. She was eager to move out of her in-laws' home where we had been living with her late husband, my father. From that point on, our household consisted of just myself, my mother and her new husband—my uncle and her previous brother-in-law. Their marriage was fraught with challenges from the very start.

It would be more than a decade before my mother and my uncle transitioned into their new roles as husband and wife, having their first child, my first sibling, after eleven years of marriage. Looking back, it was a period of profound change and adjustment for all of us.

Even before my siblings were born, our home was filled with constant arguments and fighting over the simplest things. My mother and my uncle were completely incompatible. She hated being around him, and he hated being around her.

My mother told me once that my biological father and my uncle had not seen eye to eye. Yes, they were brothers, but they were completely different people. My biological father had been an idealist. He was kind, loving, and very protective towards my mother. He was handsome and had the mind of an engineer. My uncle was very different. He was short with round features and had a thick dark mustache. His demeanor was always harsh and borderline abusive. He was a smart lawyer but never practiced law, instead he tried to hustle and create different types of businesses. Unfortunately, he didn't know how to properly run a business, and money became very tight after the first Gulf War.

My uncle was the only father I ever knew, and I grew up calling him 'dad.' However, my relationship with him was far from what I wanted it to be. I yearned for him to be a loving presence in my life, but he was more like an absent father at best. I craved his approval and affection, but I can't recall him ever telling me that he loved me or that he was proud of me. Given the traumatic circumstances of their marriage, I often wondered if he ever saw me as his son. I always felt like he resented me because I was the reason he had been forced to marry my mother. This strained dynamic cast a shadow over my childhood, leaving me to navigate most of the complexities of our unconventional family on my own.

The only thing my dad and I did bond over was our love of Western movies and TV. We'd watch Rambo and Star Wars and various commando movies together. He loved Clint Eastwood and Charles Bronson, and so did I. When the show Remington Steele with Pierce Brosnan came out, I was fascinated. Pierce was the head of a private investigative agency, and I dreamed of having a thrilling and dangerous job like that, too.

Despite the complexities of our family dynamic, I was generally a happy child. My mother poured all her love into me, as I was the only child born from the man she truly loved. Her affection towards me was unwavering, and even amid the challenges, her devotion made my early years warm and nurturing. My mother was beautiful, with long dark wavy hair. She could have been a model in the '80s with her pencil skirts and shoulder pads. Back then, many women dressed in Western styles and didn't wear a hijab, the traditional head covering for Muslim women.

My mother's side of the family spoiled me with an abundance of toys like Legos and Matchbox cars. They always made my birthday a grand celebration. My mother would cook my favorite treats and put up decorations and balloons while I was at school. By the time I got home, she would have a big surprise party ready. I felt deeply loved by all of them—except for my maternal grandfather, the diplomat. He was unapproachable, and his stern demeanor terrified me. Tall and stoic, with a thick mustache, he never showed affection to anyone.

Those on my father's side of the family showed less affection toward me, with the exception of my paternal grandparents. My grandfather taught me to play backgammon, and my grandmother always prepared delicious food for me. She even let me steal hot naan bread, fresh from the clay oven in the garden. My father's youngest sister, my aunt, helped me with my English homework when I was in elementary school. We'd sit on the big white swing in the garden, and she would patiently correct my pronunciation.

In her late teens, my aunt began attending the English Department at the College of Languages at Baghdad University, and she often listened to English-language music like George Michael's "Careless Whisper" and Sandra's "Maria Magdalena." Watching her sing along and understand English sparked my desire to learn and speak the language as well as she did. It fascinated me that people spoke languages other than my own.

Occasionally, my dad would take me to a park near the Filipino embassy. It was the late 1980s, and migrants from the Philippines had set up an open-air market there.

"Look at all this stuff," my dad said, eyes gleaming as he surveyed the bustling stalls. "Let's find something interesting."

He was always on the lookout for items he could buy and resell for his business. One day, he spotted a table loaded with electronics.

"Dad, can we check out those tape players?" I asked, pointing excitedly.

"Sure, go take a look," he replied, leading me to the table.

I begged my dad for the Sony tape player I discovered. And, after some haggling, my dad bought the tape player, along with several cassette tapes. My cousins and I gathered around the stash like treasure hunters.

"Which one do you want?" my cousin asked, rifling through the pile.

I dug through the cassettes, my fingers brushing over the colorful covers until I found one that appealed to me.

"This one!" I exclaimed, holding up a tape.

It was an album by the American band Cinderella.

"Go ahead and play it," my cousin responded.

Excited, I slipped the cassette into the tape player and hit play. As the music filled the air, I knew I'd found my favorite band. I played that album over and over, the melodies becoming the soundtrack of my childhood adventures.

When I was in fifth grade, I memorized one of their songs without having any idea what the lyrics meant. I just imitated how the words sounded. One day, my friends told our English teacher about it, and she called me to the front of the class to sing. So there I was, mumbling this English-language rock song in front of all my classmates. The class erupted in giggles, but it was that moment, standing in front of my peers, that further ignited my initial desire to speak English. Music had planted the seed, and from then on, I was determined to understand the language behind the songs I loved so much.

Elementary school was also the time of my first memory of Saddam. I had heard the stories of Saddam holding me as a baby, at the palace. But I personally remember the day my classmates and I had to parade the streets showing our loyalty.

I stood in the dusty schoolyard, shoulder to shoulder with my classmates. The morning sun was already hot, and I could feel sweat starting to trickle down my back. There was a nervous energy around us, an unspoken understanding of what was about to happen. Our teachers shouted orders, their voices sharp and impatient, herding us like cattle toward the row of buses parked outside the school gates. I climbed onto one, and found a seat next to my friend, my legs barely reaching the floor. The bus was crowded, filled with the chatter of excited, anxious voices. We were headed to the Al Mansour district, an area with wide streets and grand houses. But this wasn't a fun school trip to see the nice part of the city. Today was different.

When we arrived, they marched us off the bus and lined us up on a side road. Men from the Ba'ath Party, their faces stern and their uniforms neatly pressed, moved among us, handing out signs, flags, and palm tree leaves for us to carry. I could feel the weight of expectation pressing down on me, and I knew I'd better follow suit.

We were ushered to join dozens of other schools, a sea of children flooding onto the main street. The teachers and Ba'ath Party officials barked commands, pushing us into lines, making sure we all looked perfect. I clutched my sign with both hands, my palms sweaty and my heart pounding. As we started to march, the chants began—loud and rehearsed.

"Long live our leader Saddam! With our soul and our blood, we will sacrifice ourselves for you."

My voice joined the chorus, blending in with the others. I looked around at the adults lining the streets, their faces a strange mix of pride and something else—something like fear.

A few years later, when I was eleven, I started hearing my father and grandfather talk about another potential war. Their voices were hushed but tense, the gravity of their words unmistakable.

"This time," my father said, leaning in closer to my grandfather, "it will be much bigger than the war with Iran."

My grandfather nodded solemnly, his eyes fixed on the radio. They stayed glued to the Arabic-language station

Monte Carlo because it had more information than the state owned station controlled by President Hussein's son, Uday —the unpredictable one who kept lions and tigers as pets in the palace.

I sat nearby, pretending to read a book but listening intently to their conversation. The crackling voice on the radio filled the room with snippets of news, painting a grim picture of the unfolding events. Every mention of foreign armies and military maneuvers sent a chill down my spine.

Everyone was scared our country would be decimated. My paternal grandmother's brother had a twenty-four-acre farm about an hour south of Baghdad. And it was decided that we should all go there for a few weeks to weather the storm that seemed imminent.

When we arrived we found the farmhouse was a small light brown building with just two rooms and a bathroom. It was generally used by the local farmer my dad's uncle had hired to look after the farm. I guess that when we showed up en masse, he went home to his own family. The group included my grandmother, her children, their spouses, and their kids; her sister, her kids and their spouses; as well as, her brother, his wife and their kids. There must have been thirty-five to forty people, if not more.

It was around August, and in Baghdad, the nights were much cooler than the days during that time of year—it was usually quite pleasant. Still, we chose to sleep indoors for safety reasons, all of us in the two-bedroom house. We had to stack mattresses on the floor, placing them right next to each other to fit everyone inside, like pieces in a game of Tetris.

For the first couple of weeks, we lived off of what we had brought with us from home and bartered with neighboring farms for other things we needed. We'd brought bags of rice and flour and got meat and produce from our own farm. The rest we sourced from the nearby farming community, which went on for miles.

One night, we were awakened by the sound of explosions in the distance. The sudden, violent noise jolted us from our sleep, and panic quickly set in. The women and children began crying, clutching each other in fear, while the men hurriedly went to the rooftop to see what was happening.

My uncle came back downstairs, visibly shaken,

"It's just a web of fire and explosions over Baghdad," he said.

I was terrified. I didn't understand what was happening to my country. The year was 1990. And this was how I witnessed the beginning of the Gulf War.

Things quickly became more difficult for us on the farm. We started rationing our food because we realized it could be a long time before we could even consider going home. Hunger became a constant companion, gnawing at us day and night.

But my uncle did his best to keep our spirits up. One afternoon, he called us over with a grin.

"Come here kids, I've got something for you."

We gathered around as he held up what looked like a makeshift slingshot.

"I made these for you," he said.

He explained how he had cut rubber from an old tire inner tube and found branches with the perfect V shape. He had even attached a piece of leather to hold the stones.

"Why don't we go hunting?" he suggested, his eyes sparkling with a mixture of mischief and determination.

We followed him into the fields, our slingshots ready. My uncle showed us how to aim and shoot. Before long, we had shot a few pigeons and smaller birds. We collected our spoils and gave them to my uncle. He showed us how to pluck the feathers, gut the bird, and skewer them over a fire.

"Look at that," he said, turning the skewer slowly. "Dinner is served."

We sat around the fire, eating the freshly roasted birds, proud of our accomplishments. Each bite tasted like victory, a small triumph against the backdrop of war. My uncle tried to make it feel like an adventure, telling us stories and cracking jokes. But even as kids, we could see the heaviness in his eyes and sense the silent burden he carried. The weight of the war was never far from his mind, no matter how hard he tried to shield us from it.

The initial round of airstrikes, later known as Operation Desert Storm, didn't last long. After only a few weeks, we were able to return to Baghdad, but the war dragged on for about six months. You have to understand, when we fled to the farm, the news was telling us that the enemy intended to destroy our country. We expected relentless shelling, airstrikes, and countless civilian casualties. But that's not how it unfolded.

In hindsight, we realized all the propaganda was under Saddam's tight control. Nothing aired without his approval. The regime spread rumors that the Americans were deliberately targeting innocent children. And people believed it, panicking in fear. Yet when the airstrikes actually began, they were strategically aimed at Saddam's palaces and military installations. In a twisted move, Saddam started inviting civilians to sleep in his palaces, using them as human shields.

One day, an air raid shelter was hit, and we later learned that Saddam himself had been there, seeking refuge. He had been counting on the Western tradition of avoiding noncombatant targets, using innocent lives to protect his own.

The Gulf War finally ended in February, and afterwards, life in Baghdad became incredibly difficult. The city's infrastructure had been heavily targeted, making water and electricity scarce. On top of that, the economy was in free fall due to the United Nations Security Council's crippling sanctions which would last for thirteen years, until 2003. Though the fighting had ended, these sanctions kept the country in a state of crisis for over ten years.

In an attempt to alleviate the growing suffering, the Iraqi Minister of Commerce proposed a program called 'The Ration' which Saddam Hussein quickly approved. This program offered a monthly package of basic food items at a nominal price, distributed by the Ministry of Commerce to both Iraqi citizens and resident foreigners living in the country. The idea was to help people cope with the severe shortages caused by the economic blockade and sanctions that had been imposed.

I remember often going with my mother to collect this food ration every month. Once the food was counted and weighed, we loaded everything we had received into a wooden cart borrowed from our neighbors. At first, the ration program provided some relief, but as time passed, the quantity and quality of the food we received declined. The country was starving and growing desperate.

The oppressive sanctions created all-consuming financial stress for most Iraqis, and my father was no different. Sanctions only hurt the innocent, not Saddam nor his henchmen. They were still living large, while my father's businesses were falling apart. At one point, he wanted to sell our house—the house my mother had bought with her widow's compensation—but Mom wouldn't hear of it. And the house was in her name, so he couldn't touch it. Debt collectors would knock on the door every day, looking for my dad. This would trigger intense, sometimes physical fights between my parents over money.

Over the years, our home became even more stressful with the addition of my two sisters and my brother—three more mouths to feed. For their privacy and safety, I'll refer to them by pseudonyms: my sisters as Fatima and Amel, and my brother as Mustafa.

When I was thirteen years old, Fatima, my first sibling, entered the world, and my father declared it was time for me to start pulling my weight. I was expected to get a job. After school, I'd return home to assist him in repairing electronics for resale. Eventually, we both found employment at Saddam's Republican Palace, the very same place I'd been taken to as an infant. It was nestled along the banks of the Tigris River, and it would later become part of the Western-military controlled 'Green Zone.'

This palace held terrifying connotations for everyone in Iraq. It was a place of intense security. Ruthless guards and intelligence operatives patrolled it in unmarked cars day and night. If your car broke down near the palace, you knew it was best to leave it. Staying near the palace for too long could lead to suspicions of being a spy, which could result in detention and even torture during interrogation. Just driving nearby raised intense fear in all of us, and that's where we went to work.

One of my jobs was connecting chandelier crystals, a painstaking task that left my fingers sore and bloody. Each time I assembled the crystals with those tiny metal connectors, they pricked my fingers, turning the delicate work into a painful ordeal. But that's what we did; we had no choice.

We worked under difficult conditions as laborers in construction and remodeling. We were monitored constantly and sometimes were forced to stay at work late into the night. It was physically and mentally draining work, but this was the only way we could survive.

2

DREAMING OF A FREE IRAQ

Within just a couple of years, when I was about eighteen, family connections opened a better door for me. My aunt's husband was friends with an older gentleman who held a prominent position in the Iraqi intelligence community. And since I was already very skilled in English at that point, he agreed to help me apply to the College of Languages at Baghdad University. I was excited by the idea of following in my aunt's footsteps.

The older gentleman and I walked into the college dean's office. My heart pounded with anticipation as I clutched a folder containing my application paperwork. The gentleman had advised me to agree with whatever he said, and I nodded, ready to follow his lead.

The dean, a stern-looking man gestured for us to sit.

"Good afternoon," he said, glancing at my folder. "Let's get started. Tell me about your high school scores."

I took a deep breath.

"I did well, sir," I replied, listing my scores as confidently as I could.

The dean listened, then picked up his pen. I watched as he scribbled something in green ink on my folder. This was the signal we had been hoping for. Green ink was used by a higher official when everything was good to go.

The older gentleman smiled and gave a subtle nod.

"Thank you, sir," he said warmly. "We appreciate your time."

The dean nodded, closing the folder.

I could hardly believe it. That was it. Relief washed over me. Within three days, I would be starting night classes, ready to embark on this new chapter of my life.

The first night of classes, the professor walked in and introduced himself in Arabic. Then told us what we were to expect from his class.

"This is the last time we will speak in Arabic," he said "I encourage you to participate in conversations, ask questions, and make mistakes. You will learn from them."

From that moment forward, all of my university education was in English.

During my first year of college, I struggled to find my place and the right group of friends. But eventually, I found my clique. The College of Languages taught many different languages, so I developed friendships with students who were learning Russian, Spanish, French, Turkish, Hebrew— all kinds of languages. My two best friends were Khalid, who was studying Russian, and Najar, who was studying Spanish.

Khalid, was a tall man with light, curly hair and piercing blue eyes. He had a distinctive shuffle in his walk, and his long arms were reminiscent of a basketball player.

His calm demeanor and idealistic nature were always evident.

Najar in contrast, was short with tan skin and dark hair. He was easy to talk to and often expressed his strong opinions with a touch of sarcasm. He loved to party and drink, a pastime his family accepted given that they were agnostic, not Muslim.

We all connected because we loved different languages and cultures but also because we loved listening to the same Western music. These are the friends that would eventually become interpreters with me—or 'terps' as the military called us. We became the three musketeers and were inseparable, starting in our college years.

We would walk along the palm tree-lined campus and hang out together at the snack shop. Here they served only non-alcoholic beverages, of course, since almost all the students were Muslim. Its pork-free food options included hot dogs, burgers, fries, pizza, and chips. But what was served on campus didn't limit what we would eat or drink. Nor did we let the campus' Ba'ath party leadership define what we would think. The Ba'ath Party, under Saddam, had evolved from a multinational Arab unification movement to a highly centralized authoritarian force in Iraq. Party membership became a prerequisite for government jobs and university admission.

However, it was important for translators to understand the cultures related to the language they were studying. And, thus, college allowed us more freedom to engage in elements of Western culture. We were at school not just to learn a language but to be exposed to the culture of people who spoke it, which was somewhat controversial. American ideals didn't really mesh with Iraqi culture. We were supposed to learn the culture, know the culture, but never become part of it or believe in it.

Call me a rebel, but all I wanted to do was create my own cultural norms influenced by a plethora of cultures. I think most college kids go through this phase. And while music was my entrée into developing this new paradigm, college gave me the freedom to develop it further.

The whole college culture was immersed in Western music. We'd listen to music while playing basketball or just hanging out smoking. My passion for Western music skyrocketed by my second year. I went from having five or six favorite bands to having fifty or sixty, thanks to Najar and Khalid. I'd write down lyrics and try to translate them to better understand the language. I remember going to one party where a Metallica cover band from Iraq was playing. It was the first time I heard the song "Fade to Black," which would become a strangely prophetic anthem over my life. A tiny dark seed took root in my soul that night, which over time would slowly begin to grow.

My goal for going to college was to prepare for a good job so I could reduce the financial burden my parents carried. But I was the worst student ever. I would show up on campus when the shuttle—a 22-seat Toyota Coaster bus—arrived around 3 p.m., then play basketball for a while, maybe attend a class or two, then head over in Najar's car to a liquor store and buy gin, vodka, or cigarettes. We'd then go to Radio One, a music shop that carried only English language music and hang out drinking and smoking in the music store. I'd never had access to alcohol before, but my friends who were financially better off would share with me. When it was getting dark, we'd head back to college in order to make the 8 p.m. shuttle.

I worked a few odd jobs during this time like at the print shop, to help pay my tuition. But I also always hustled. In my third year, Khalid and I decided we didn't like the shuttle services that were available, and we were tired of paying for our fare. So we went to the bus station and negotiated with a driver to take on a new schedule. We agreed on a price and then got as many students as possible to be paid riders. That got us a better schedule and, as the service administrators, we allowed ourselves to ride for free.

At the beginning of college, my attitude toward politics was neutral. I'd lived my entire life under a dictator and that was all I knew. Most people didn't like Saddam, but we were loyal to him out of fear. And well, what was the alternative?

One afternoon I was walking across campus wearing a T-shirt from the 1998 Olympics. It had the Olympic five-rings logo on it and also a US flag. A fellow student, who was a member of the ruling Ba'ath party, approached me.

"Hey, this is unacceptable," he told me. "Why are you wearing the enemy's flag?"

"What? This is the Olympics!" I answered. "Are you a student? Do you understand what the Olympics is about?"

He grabbed my shirt and tried to rip it off me. Luckily, Khalid got between us and calmed things down. Later, as we rode the shuttle home, Khalid asked me about it.

"Why didn't you tell him your father was a patriotic martyr in the war?"

That's when I began to develop my deep hatred for the Ba'ath party and Saddam. I realized they were always watching. And after that, I got to know really quickly which students in our classes were members of the Iraqi Intelligence and which student union reps reported directly to Saddam's son Uday—who headed the brutal Fedayeen, Saddam's paramilitary organization.

They controlled every institution in the country. You couldn't get a job if you were not part of the Ba'ath party; your social credit revolved entirely around that. It was a kind of 'if you're not with us, you're against us' way of thinking. Since the Ba'ath party had, by this time, become predominantly Sunni, this was also creating divisions along religious lines in our country.

In Iraq, all males were required to serve in the military for three years upon college graduation. I was just about to finish my third year of college and was feeling the looming dread of mandatory military service. I couldn't stomach the possibility of dying in a war to protect a dictator I had begun to hate.

Unbeknownst to me, my mom and dad had started discussing possible ways to keep me out of the army. My mom was adamant on getting me out of serving since my father had been killed in the war. The only solution they had figured out was to use the Badel program, which allowed you to pay a fee to avoid high risk military assignments. If you paid $1 million dinars, the equivalent of about $1,000 US dollars, you could get 'flag service,' just three months of safe duties. But my dad was so deep in debt that this wouldn't be a feasible solution.

So that year, I came up with my own solution. As I sat at my desk about to take my final exam, I stared at the blank sheet in front of me, weighing my options. I knew the answers to the questions, but proving that wasn't the point; this had become about something much bigger. Failing the exam was the only way I could stay in college, the only way to buy myself more time.

With a steady hand, I wrote my name at the top of the page and then set my pen down. My heart pounded in my chest as I looked around at the other students, all of them hunched over their desks, focused and determined. I felt a twinge of guilt, but it was quickly replaced by a sense of resolve. I'd made my decision.

As the minutes ticked by, I didn't write a single answer. I just sat there, staring at the empty lines on the page. Finally, the professor called time. I gathered my things, walked up to the front of the room, and handed in the blank sheet. A brief, curious glance from the professor—and then I was out the door. I didn't necessarily want to fail, but it was the only way to ensure I could stay in college for at least another two years. My plan worked.

★ ★ ★

My friends and I had all hoped to become professional translators—in our last year in college, we were already skilled enough to translate entire books. But as graduation approached, we started hearing rumors of a possible invasion. With each passing day, my dream of becoming a translator felt more fragile, like a bubble on the verge of bursting.

Everyone was worried about the economy. But no one dared to say anything negative about Saddam. The shadow of the regime loomed over every conversation, over every whisper.

"Why does the western world think we have weapons of mass destruction?"

"Why does Bush call us the axis of evil?" People confided in hushed tones.

The thought of my country being occupied was terrifying, and the uncertainty of our futures weighed heavily on us all.

As rumors of an invasion became more frequent, the tension in the air was palpable. One evening, my English professor began to reflect on the nation's situation. The room, which had been filled with the hum of idle chatter, fell silent as he spoke. His voice carried a weight of concern that none of us had heard before.

"You all have been watching the news," he began. "And as you all know, our country might get attacked."

The gravity of his words hung in the air, heavy and foreboding.

"It's an atrocity that we are in this situation again. Everybody is on the wrong side."

We shifted uncomfortably in our seats, the reality of his statement settling in. The thought of conflict looming over our heads was almost surreal. Yet here we were, facing the possibility of a war once more.

"Be safe, all of you. Be careful as we go into this next season. We don't know what things may come. The future of our country is unknown."

His eyes scanned the room, meeting each of ours in turn. The uncertainty in his gaze mirrored our own fears.

"Don't talk, don't say too much, don't engage," he continued, his tone growing more urgent. "Any decision you make will have a big impact on you and your country."

The classroom, usually a haven of learning and discussion, felt transformed into a place of solemn contemplation. His advice was clear: caution and prudence were our best defenses in these uncertain times.

As he finished his cryptic speech, the silence in the room was profound. We were afraid to say anything because we didn't know who might be watching or listening to us.

So we sat very still the remainder of the lesson, our fears holding us rigidly in place as we tried to imagine what lay ahead.

"What changes might be coming? Would this occupation bring democracy? What could Iraq become if Saddam was no longer in power?"

And maybe others were also thinking as pragmatically as I was.

"Would there be any jobs available if we end up at war? How would the war impact my future?"

These questions swirled in my mind, each one heavier than the last.

I think my professor's riddles were his attempt at preparing us for what was to come. He urged us to think beyond the surface, to consider the deeper implications of the times we were living in because he knew the gravity of what was about to happen.

The US invaded Southern Iraq in March 2003, starting near Basra where our oil production was centered. My college friend Lu'ay, from the Spanish department, had been drafted. And, because he wasn't related to anyone important, he got put on the front lines. But as soon as the coalition invasion began, he came back to Baghdad.

"I'm not going to sit there and die for nothing. The officers had abandoned their posts. There was no ammunition and very few weapons. I didn't have anything to fight with," he said.

That was a story from the frontlines that we weren't going to hear on the state-owned TV or radio channel. Saddam controlled the invasion narrative, and we would get daily briefings from the Information Minister, nicknamed 'Baghdad Bob' by the US. He would report that American soldiers were committing suicide 'by the hundreds' and deny that there were any American tanks near Baghdad. In reality, tanks were rolling in as he spoke, and you could hear bombs in the background of the broadcast.

Within just a few weeks, US troops had moved across the country towards Baghdad. It was an eerie time. The Fedayyin, translated 'Self-Sacrificer,' and Ba'ath party members were sandbagging street corners as emplacements for AKs, the Soviet assault rifles, and for RPGs, the rocket-propelled grenades, trying to fortify the city.

There was a lot of chaos and confusion. We had food, oil, and gas shortages. There were huge lines of cars at the gas stations because there was such low supply—no new gas was coming in. No oil was being refined. Most of our oil came from refineries in the south. And those were cut off even before the fall of Baghdad, which was of course strategic: oil was what funded Saddam and the country.

A chilling rumor circulated, claiming that fifteen or sixteen high-ranking military advisors had dared to speak truth to Saddam, informing him of the nation's bleak prospects in the war and predicting Baghdad's imminent fall within months. In response, Saddam swiftly orchestrated their execution, eliminating any dissenting voices. Yet, in a twisted display of propaganda, the media depicted this as a scene of apparent unity—a massive gathering of hundreds of officers applauding and cheering in a grand hall, with Saddam at the center. This facade, carefully constructed by Saddam, was emblematic of his mastery in manipulating the people's perception—a skill that had enabled him to wield power for almost thirty years.

For two weeks, we heard the airstrikes against the palace, just a little over a mile from where we lived. Bomb blasts shattered our windows and shook our dishes, the sound reverberating through our home. I remember climbing up onto the flat roof of our house to watch the explosions. The night sky was lit up with fiery bursts, and the ground trembled beneath me. I felt sick to my stomach, conflicted and horrified at the thought of innocent people dying. Yet, at the same time, a part of me questioned, maybe this is the only way to get rid of Saddam?

The first few weeks of the war were very confusing. We didn't understand what was happening. We tuned into Baghdad Bob and heard one story.

"The American's have reached this location. We defeated them."

Then he'd announce that they'd reached another location, and that they'd been defeated again. The dissonance between Baghdad Bob's proclamations of victory and the reality of bombs falling around us was jarring. We knew better than to believe the official line, but the truth was elusive, buried beneath layers of propaganda and fear. The uncertainty gnawed at us, making the days feel interminable and filling the nights with unease.

In those moments, music became my sanctuary. Guns N' Roses blared through my headphones, drowning out the sounds of war. The raw energy of their songs transported me to a place where the chaos of my surroundings faded, even if only temporarily. It was my coping mechanism, my way of holding onto some semblance of normalcy amid the turmoil.

The war was omnipresent, but inside my bubble of music, I could pretend it was all a bad dream. Here, my thoughts could drift to the mundane worries of a college student: exams, friendships, and unspoken crushes. It was a stark contrast to the reality outside, but it was all I had to keep myself grounded.

Looking back, I'm amazed that, with the whole country in shambles, I was just waiting for this war to be resolved so I could go back to that simple world of mine. All I cared about at that moment was myself—I was a typical self-centered college kid.

As the conflict dragged on, the weight of the situation began to sink in. The future I had envisioned—one filled with possibilities and new experiences—seemed increasingly out of reach. The uncertainty about whether I would be able to complete my studies, pursue a career, or even see my loved ones again was a constant source of anxiety.

One day I heard my dad and mom talking with my grandparents. My mom's dad lived near the college, northeast of us, further away from the palace and the explosions. He wanted us to come stay with him.

"I'll come get you," my grandfather told them. "You need to come to my house. It's getting dangerous where you guys are at."

My parents didn't agree right away, but with tensions mounting, they finally arranged for my grandfather to pick us up the next day. We didn't own a car, and finding a ride would have been impossible since we were in a war zone.

Late afternoon the following day, we spotted my grandfather's truck turning onto our street. I felt a moment of relief as we quickly loaded our belongings into the back of his truck and squeezed my entire family onto the truck cab's bench seat.

After a while, we pulled up to a stoplight at an intersection that led onto an overpass. This would lead us across the Baghdad Airport Highway, which the Americans would later rename 'Route Irish.' Looking down, I noticed a police car at the highway entrance—and sandbags piled up on the highway below the overpass. Republican Guards and members of the Ba'ath Party were waiting for the Americans. Just as we were about to get onto the overpass, everything seemed to stop. We began to hear the rumble of tanks on the pavement, a sound that made my blood run cold.

My grandfather began debating whether we should continue or turn back, when suddenly we heard a big explosion. *Boom.* The police car had been hit by a shell fired by a tank. Upon impact, the car flipped in the air and caught on fire as it hit the ground. My grandfather slammed on the gas, and we sped up the ramp towards the bridge. Just as we reached the center of the bridge, I opened my eyes for a split second and saw the longest convoy of tanks, engaging in a firefight on the highway below us. I held Fatima and Mustafa down while my mom held Amel. My dad shielded my mother as bullets began to fly over the truck. *Bang. Bang. Bang.*

We finally made it across, but I was terrified. My heart was beating out of my chest, and I was gasping for air. I didn't calm down until we reached my grandfather's house thirty minutes later.

Moments after we arrived, we heard more loud explosions in the distance and a very strange sound. *Brrrrrrrrrrrrrp.* These were the A10 aircrafts flying overhead. These started firing large caliber rounds in the direction of Baghdad. Amidst the explosions echoing over Baghdad and our narrow escape from death, I found myself grappling with new haunting questions.

"Would we survive this war, and if we did would we have a home to return to?"

★ ★ ★

Within a month, Baghdad had fallen, and rumors of looting began to swirl. Fearing we might lose our home entirely, my parents decided to return to it. However, what awaited us upon our arrival was far from the familiar—our neighborhood had plunged into chaos and confusion.

The schools and universities were closed, their gates locked and classrooms abandoned. US soldiers and Humvees patrolled the streets, with military personnel operating checkpoints all around the palace, controlling access to it. At US military checkpoints, some people were allowed in while others were turned away. Where entry was granted, chaos ensued as looters took whatever they could find—computers, furniture, rugs, chandeliers, anything that could be sold.

Rumors spread like wildfire of discovered gold and other treasures, enticing even more to join the plunder. Before the war, the palaces were opulent residences; they served as a resort for high-ranking officials and Ba'ath party members along with their concubines.

It seemed like the initial goal of the coalition forces—consisting of the United States, United Kingdom, Australia, and Poland—was to clear out the palaces for their own use. The looting was rampant, a stark contrast to the order we had once known.

As the looting escalated, some Iraqi civilians took matters into their own hands. Some set up makeshift checkpoints, urging people to do what was right and cease the looting. Others saw the plunder as retribution, believing that everything within the palace walls had been stolen from the impoverished Iraqi people and should be reclaimed.

This marked the first significant ideological divide in the aftermath of the invasion.

Since the fall of Baghdad, nobody really knew what was going on. Baghdad Bob was gone, and the only news we had access to was the local radio station now run by the coalition forces. Their messages were more disturbing than inspiring. And with our history of controlled news, we questioned what we heard.

"We're here to help," they would say. "Saddam is gone."

Then they'd ask anyone with information about Ba'ath party members to come and talk with them. This was both different and strangely familiar. We were accustomed to the notion of informants lurking around every corner; now they were simply expected to report on a different group of individuals.

The first month following the invasion was a blur, a whirlwind of uncertainty and chaos that extended far beyond the military maneuvers. The 'shock and awe' we endured transcended mere military action, leaving an indelible mark on our collective psyche. We weren't sure what to believe. Was Saddam truly gone? Throughout the neighborhood, people congregated outside, watching and whispering.

One day, one of my college classmates Rami dropped by, and we eagerly asked him what he'd witnessed on his car ride over to us, hoping for some real news.

"People were looting everywhere," Rami reported, his eyes wide with astonishment. "And I saw US Army checkpoints, too."

In Baghdad, armed figures had typically been associated with the police or the Ba'ath party; the sight of American soldiers wielding weapons felt foreign and unsettling. The rapid transition from tanks rolling in to Saddam's downfall had left us all in shock.

"I'm trying to make sense of it all," Rami confessed, his voice tinged with panic and perhaps a hint of excitement.

I, too, was still trying to make sense of it all. But I knew it was true—Saddam had been deposed. And, while the city was plunged into disorder, the faint hope of freedom began to take root. I couldn't help but ponder the new possibilities.

"Could a free Iraq truly be within reach? Was this the beginning of a new chapter for my country where life might finally improve and we could have a democracy instead of a dictator? Would the US occupation liberate us?"

As a naive college student, I found myself yearning to be a part of rebuilding and reshaping a new, free Iraq. And what better way to contribute, I thought, than by utilizing my language skills and knowledge of Western culture.

3

TERP OR TRAITOR?

As the eldest son, I felt a huge sense of responsibility to find a job and help support my family. As you can imagine, jobs were hard to come by because of the war. One day, as I navigated through the bustling open-air market teeming with stalls selling fresh fruits and vegetables, I noticed a US Humvee parked in the middle of the road and saw an opportunity. It felt serendipitous; maybe, this would be the break I had been hoping for. Perhaps, I could alleviate the financial strain on my family, finally make my father proud, and also contribute to rebuilding a free Iraq.

With a mixture of anticipation and nervousness, I approached the Humvee and looked up at the man sitting on top of the vehicle, with his hands on the machine gun. Technically, I still had a few months remaining until graduation, but since my college was closed indefinitely, I figured it was now or never.

"Hey, do you need translators?" I called out.

He gave me a bit of a hard time.

"Can you speak English?" he asked, raising an eyebrow.

"I'm talking to you, aren't I?" I snapped back.

The soldier paused.

"Alright, fair enough."

He directed me across the street to where his boss CPT Smith was talking with a few locals. He had dark skin but European features. Today, I'd figure maybe he was creole, from Louisiana. At that early stage of the war after Saddam was deposed, the US goal was to rebuild the country, lift sanctions, and ultimately develop a democratic government in a society deeply rooted in tribal systems. As part of the initial strategy, soldiers were stationed in public places like markets to build relationships with Baghdad natives. Their goal was to identify key contacts in the community and gather general intelligence about the area.

This was my first interaction with an American military person. And as CPT Smith stood before me, I noticed his uniform was crisp and imposing.

A bit flustered, I asked again.

"Sir, do you need translators?"

"Yes. But why do you want to be a translator?" he pressed.

"My family needs the money," I replied earnestly. "And I want to be part of rebuilding the country."

He nodded, appreciating my response.

"Good. Well, we're here to help with that."

Then he turned to his interpreter.

"Give him more information about the gig," he instructed.

His terp gave me more information on where to meet the Americans. I managed to meet them two days later outside of a clinic. Another hopeful interpreter, Bashar was also there.

We both stood nervously as an American soldier evaluated us. After a few questions, he nodded.

"You're both hired," he announced. "And you're going to work with Civil Affairs."

I was ecstatic, but also curious.

"How much will we be paid?" I inquired.

"It's five dollars a day," the officer replied briskly. "Do you want the gig or not?"

Five dollars seemed like a fortune to me in 2003, given the sanctions and all the financial hardships we had endured.

"Yes!" I replied emphatically. "I want the job."

Looking back, it's hard to believe we risked our lives for such a small amount. But these were desperate times, and five dollars a day felt like a lifeline in a world turned upside down.

At first, we were not given uniforms, body armor, or even helmets. We just wore our civilian clothes and something to conceal our faces like a scarf or gaiter. We made sure not to tell anyone we were terps because a lot of the locals would have thought we were traitors. And as I started to think about my cover story and my strategy for getting back and forth to base without getting caught, I suddenly felt like Remington Steele. My dream of a dangerous job was coming true.

In 2003, I joined a five-soldier Civil Affairs team led by CPT Smith, part of US Special Operations, stationed at

Camp Slayer. Civil Affairs' primary goal was to forge connections with local leaders and tribal bosses in order to identify community needs and address infrastructure issues that the Army could help reconstruct or repair in collaboration with local Iraqi contractors.

Camp Slayer resembled a luxurious resort with numerous buildings and houses formerly used by Saddam's guards or concubines. There were small man-made lakes where soldiers enjoyed recreational fishing. The main house had a beautiful wooden double-door entrance leading to a spacious round hall with marble floors and top-quality finishes. Rooms flanked either side of the hall, where the US team had already set up folding tables for their equipment and computers. I remember my first day, walking into this house, meeting the team, and seeing my very first DVD player. One of the soldiers was watching *Band of Brothers*, a show I would later come to love.

The translator office at Camp Slayer was simple, just one long room with a satellite TV, a desk, and some couches. We started with an incredible group of about fifteen translators. We all came from humble beginnings.

Bashar was the first translator I met. His upbringing was unique, having spent much of his life in New York City because his father worked for the Iraqi embassy. He was more familiar with African American culture than Iraqi culture which made him stand out. But Bashar was a natural leader and understood American culture better than the rest of us, so he was assigned as the lead interpreter, responsible for scheduling all of us. His easy going demeanor, always joking and laughing, endeared him to everyone, including all of the American soldiers. Bashar made friends effortlessly, bridging the gap between cultures with his charm and humor.

Sa'ad was one of the translators recruited from the local population, most of whom had learned English through watching TV. He had a stocky build with tan skin and short, dark hair. His square-shaped head was accented by a short walrus mustache, giving him a distinctive look. He was a quiet presence among us, often keeping to himself. From the start, there was something about him that made it hard to connect with him. While he would occasionally joke around with the other translators and offer help when needed, there was always a sense of distance.

A lot of people didn't feel comfortable around him, though they couldn't quite pinpoint why. He wasn't standoffish, but he didn't seem to fit in seamlessly with the group. As a result, most of us didn't get to know him very well. Our personalities just didn't click, and that unspoken barrier kept him somewhat isolated, even within the same camp. Sa'ad bonded mostly with two other translators, Ethar and Ahmed, who were his neighbors. The three of them formed a close-knit trio and were often seen together during breaks and off-duty hours. Ethar and Ahmed seemed to understand Sa'ad in a way the rest of us didn't, sharing inside jokes and a camaraderie that highlighted the divide between the self-taught translators and the college-educated ones.

Ethar was a younger guy with a good head on his shoulders—maybe eighteen or nineteen—so young that, when we started working together, he didn't have any facial hair yet. He'd been hired locally and had a lot of potential despite his mid-level English skills. He had lighter skin, a slight build, and a serious demeanor that contrasted with his age. Ethar's dedication and eagerness to learn made him stand out among the other locally hired translators.

Over time, several guys named Ahmed worked on our team. Ahmed is a really common name in Iraq, so we called Sa'ad's neighbor 'Skinny Ahmed.' He was also young —only sixteen when he was hired—and close to six feet tall, scrawny like lots of boys who've grown tall really fast. And he was also too young for facial hair. Like other young translators, he gradually grew into the position, learning and adapting quickly despite his age. The presence of these young men, especially those like Ethar and Skinny Ahmed, added a unique dynamic to our team. Their youthful energy and determination were both inspiring and essential in the challenging environment we operated in.

The first week on the job, I was given the assignment to translate between Civil Affairs, local tribe leaders, and village elders on a very sensitive case. Civil Affairs had informed the local leaders that there were bodies of fallen soldiers scattered around the airport.

This was a delicate task and I had to be sensitive when translating to the local tribe leaders. On one hand, it showed that the US was trying to approach the situation with respect by allowing Iraqi families to properly bury their soldiers killed in action. But from the locals' perspective, it felt like the enemy was admitting responsibility for murdering their countrymen, leaving people uncertain how to respond.

In Muslim tradition, it's very important to bury the dead right away, so this couldn't be delayed. That same day, the bodies had to be identified by a few elders, tribal leaders and a representative from the US-established neighborhood council. Then a truck had to carry the corpses back to a mosque to be bathed with rose water and wrapped in white linens for proper burial.

We arrived at the airport, and approached the first body, the sight of it shook me to my core. I couldn't bear to look at it, my mind recoiling in shock and terror. This was one of many. There was a body here, a body there. Some were simply lifeless; others were gruesomely mangled. The smell was overpowering, making me feel sick to my stomach.

I looked at the Civil Affairs guys.

"You're capable of doing this?" I asked myself. "I thought I knew you…"

SGT 'T' noticed my distress. She was an African-American woman with a keen sense of empathy. She gently pulled me aside, handed me a blanket.

"You just sit here," she said, inviting me to sit down.

The team continued their grim task, moving from one body to the next. I stayed back, unable to bring myself to approach the corpses again.

Looking back, that first mission was a crucible, forging a new part of my identity at twenty-three years old. It was a stark introduction to the path I had chosen, one that would shape me in ways I had yet to understand.

My first combat experience was a few weeks later on Route Irish. This was a seven and a half mile stretch of road from the Green Zone to the airport which became the world's most dangerous road. It would later be nicknamed 'IED Alley' for all the Improvised Explosive Devices that would be found here.

That day, I was riding with a few guys from Civil Affairs in a Humvee down Route Irish toward Camp Slayer, returning from an assignment. Back then, some of the Humvees didn't even have plastic doors, so it felt like riding in a Jeep with the doors off. I was sitting on the right side behind the Captain, my legs dangling outside, enjoying the ride. As we passed under the highway bridge, I saw a green Toyota Corolla right in front of us trying to merge onto Route Irish. Just after it had merged, an IED detonated just off the side of the road, hitting the Corolla instead of our Humvee. The vehicle's windows shattered. The blast was deafening, and the shockwave rocked our vehicle. We were incredibly lucky that day, saved by a split second and the presence of that unlucky Corolla. Looking back, I would count that as the first divine intervention in my life.

We pulled over immediately and tried to make sense of the situation. CPT Smith ran toward the car, and I ran after him. I had no idea what was going on military-wise; I just wanted to know what he wanted me to do.

"SGT T!" he shouted, "Call it in! We need a medevac. There's multiple wounded, including a baby."

Then he turned to me and asked me to translate.

"What's your name?" I asked the guy in the driver's seat.

Blood was everywhere, and the man couldn't move. The older lady in the passenger seat had been knocked out of the vehicle by the impact. He told us she was his mother. They were on their way home from the hospital, with his wife and baby who had just been born.

"Is my wife okay?" he asked anxiously. "Is my baby okay?"

He still couldn't move, so he couldn't see what had happened in the backseat. But I could. His wife had been impaled by glass; the baby's face was also covered with shards. This was my first view of a war casualty.

"Oh shit," I remember thinking. "This is real." I stepped back in horror.

Within seconds, we got the injured family out of what remained of their car, applied first aid, and maintained security while waiting for medevac. CPT Smith had me translating as we waited.

This IED wasn't as strong or as sophisticated as the ones that would be used later. We had no way of knowing if the IED was intended for our military vehicle or if the purpose was to observe how the military would respond.

The highway exit where the IED hit was near a residential neighborhood, and CPT Smith noticed that someone with a video camera appeared to be filming us. We ran towards the cameraman.

"Tell him to come here right now or I'll shoot!" he instructed me. "Tell him not to run."

As I got closer to him, I recognized the man. His name was Hani; we used to hangout listening to rock cover bands.

"That's my friend!" I called back to CPT Smith. "I know him. Don't shoot!"

When I was close enough to be heard, I called out.

"Hey, It's me! Waleed!"

Hani wasn't scared or even rattled. He just continued filming, whipping out a badge at the same time.

It turned out that Hani was a translator and a local reporter working for ABC News.

"This permits me to film any incidents," he told CPT Smith in English. "You can call and verify. Can I interview you?" So it's possible my first combat experience ended up on the US evening news.

After days like that, I went back to a family home that was nearly as uncomfortable as my work. My mom and I always got into contentious discussions, mostly about my future.

"I want to see grand-babies!" she would exclaim, urging me to marry a particular cousin she favored.

"No. I'm not doing it!" I'd return. Just as all roads lead to Rome, every conversation led to that.

I finally mustered up enough courage to tell her about my job.

"Yumma, I got a job as an interpreter…"

Before I could finish she jumped in.

"What? With the Americans? You're going to get yourself killed. You better quit."

"Yumma. I'm just pushing paper. I just translate documents, and I'm making five dollars a day."

I handed her a handful of dinar. Of course, I was lying to her. What was I supposed to say? I'm doing patrols, and an IED almost blew me up?

That didn't ease her mind. She remained focused on one very traditional approach to creating stability for our family. I should marry a nice Iraqi girl. Preferably a relative. Specifically, her favored cousin.

In Iraq, where families stick closely together over generations, it's important to make sure that the new bride fits in with the family. One easy way to do that is to make sure the son marries someone everyone knows well. That's usually going to be a cousin. Marrying a relative also keeps the extended family tight and, when you're part of a tribal community, it keeps the tribe together.

★ ★ ★

Four to six months after the invasion, my college finally reopened, and CPT Smith let me go back to school to finish my degree. By the time the end of the semester came around, I realized working as a terp was the best possible preparation for my final exams. By then, passing my exams was my only goal at school. My final project was in linguistics. I sat down with my professor to discuss the goals of the project I was planning. I met him in a tiny classroom, and I gave him a few pages of my draft text.

As soon as I began to explain my project, his eyes rose from the papers, and he looked at me intently.

"You need to be careful," he said. "I know who you're working with. I can see it. Your fluency is excellent. But what you're doing is very dangerous. I can't change your mind. You just need to be careful."

I just looked at him dumbfounded. I didn't know if he was threatening me or trying to give me advice.

During my earlier years at college, before the invasion, I always knew there were both spies and allies on campus, and I just kept my head down. But this was the first time I'd engaged in this kind of conversation. I walked away feeling uncomfortable. Did I now have a target on my back?

★ ★ ★

Of course, part of the reason CPT Smith let me return to college was so I could help recruit more well-qualified terps from among my friends. Khalid and Najar were really curious about my job once they saw I had a new car. One of the soldiers, SGT Jimenez, had given me an old power generator that I had then sold. And with that money I had bought a silver 1993 Opel Astra hatchback. I had never driven a car, and I didn't have a driver's license. So I'd asked my friend Rami, who had a car, to drive the ten-year-old Opel off the lot for me. He had helped me find the car in the first place. Rami couldn't believe I'd bought a car I couldn't drive.

"I'll figure it out," I told him. And I did.

Khalid and Najar had also heard about another friend who was working for the Americans in Mosul, which made them even more curious about the possibility of becoming a translator. They agreed to let me introduce them to my Point of Contact (POC) CPT Smith at Camp Slayer.

I brought Khalid and Najar to meet CPT Smith at Camp Slayer. From there, he took us in one of the big military trucks to the Green Zone where we met with Titan, the new military contracting company handling the hiring of all civilian translators. Once we arrived, they did their screenings, took their pictures and biometrics, and issued them badges. The process was surprisingly quick, and soon enough, the entire staff of translators was filled with locals who spoke decent English and my college buddies who were professionally trained translators.

When we first came on board with Titan, all their gear had the AT&T logo on it, which—for those in the know—marked us as 'terps.' Over time, different contractors took over, and the pay rates for translators fluctuated significantly. When I started, I earned just five dollars a day. Overtime, Titan dramatically increased our pay, and by the end, I was making $1,250 a month, which was a bit more than $40 a day. However, some of the soldiers mentioned that the Army was paying the contractor $6,000 a month for each of us, so it was clear who was really making the big bucks.

I was just grateful to be earning this kind of income to help my family. I even developed some pretty awesome side hustles to bring in even more money. The soldiers kept asking us to buy items in the local markets, and we always made a commission on it. We would help the soldiers buy things like gold, food, and supplements. We were making a good living. It might not have seemed like much from the standpoint of the US personnel, but for us, it was a lot.

In Bagdad, curfews had become the new normal. Once the sun was down, no one was allowed on the streets apart from Iraqi police and the military. One evening, well past curfew, we were jolted by the sound of an explosion. My mother and I ran up to the roof to see what was happening. Near the entrance to Camp Slayer, at the gate closest to the translators' office, a huge ball of fire lit up the night sky.

"Oh my God!" I said. "That's where I work."

"That's it," Mom said. "You're done. You're not going to work anymore."

I panicked a bit and started spitting out a bunch of made-up reasons she shouldn't be worried.

"They've got it under control," I said. "There's nothing to worry about. I only translate paper. It's a desk job."

"It's dangerous! Look," she said, pointing at the sky. "I'm not at peace with you doing this anymore."

Eventually, sometime after we'd gone back downstairs, I spoke out my true thoughts.

"Yumma, I can't quit. We need the money, and I kind of like this job."

The next day, I found out that the explosion was part of the controlled demolition of a concrete building that the Army didn't need. I thought I was lying to my mom when telling her that everything was under control. But it turned out I had actually been, unknowingly, telling the truth.

★ ★ ★

Missions at work made for long days. One day we were out for eight hours translating, figuring out where the local irrigation system needed repairs. By the time we got back to the team house, I was starving. My coworker dug through some boxes of MREs, the US Army 'Meals Ready to Eat' combat rations, and found a can of tinned meat.

"Here's a can of Vienna sausages," he said and tossed it to me.

I popped the top and pulled out one of the sausages. But no sooner had I taken the first bite when he interrupted me.

"Oh wait!" he exclaimed. "Do you eat pork?"

I cursed and threw the can. I completely forgot to ask if they have pork in them.

"Why didn't you wait till I finished?" I berated him.

If I had eaten it without knowing, then there would be no guilt, no shame. But once I knew it contained pork, Islam forbade me to eat it. I was starving.

When I was really young, I would have considered my family Muslims more by culture than by commitment. We were what you might call 'Ramadan Muslims.' We didn't always pray five times a day, but we observed the month-long tradition of Ramadan, which consists of fasting during the day and eating after the sun goes down.

But as I grew older, my family became more devout. Allah and the concept of his wrath or judgment was embedded into a lot of our culture. I would try to remember to pray five times a day and recite the Qur'an. But there were always a lot of rules. I definitely feared Allah and didn't want to piss him off.

Before the war, the prominent religion in Iraq was Islam, but Saddam was tolerant of other religions and also of the different branches of Islam. He and all his government leaders were Sunni Muslim, which was the minority. But Shia Muslims also lived with Sunnis in the same neighborhoods.

My family is Sunni, but we had neighbors who were Orthodox Christians. We also had neighbors who were Yazdi, an ancient mystical religion, and we all played together as children. The modern extremist ideology wasn't prevalent when I was growing up. In fact, before the war the atmosphere of the entire city changed during Ramadan. After you'd break your fast in the evening, you'd go out for ice cream and hang out with your neighbors, regardless of their sect. It was a joyous time.

A few months into my time at Camp Slayer, I was sent out on patrol during Ramadan. My task was to listen to the Khutbah, or sermon, that was being broadcast to a particular neighborhood. Mosques had loudspeakers on top of the minaret so people could hear the call to prayer, Qur'an readings, sermons, and prayers from wherever they were.

The US Army needed us translators to listen to the broadcasts because some of the imams were preaching jihad, or holy war, and other problematic ideas. So we'd sit outdoors at some distance from the mosque, listen to the service, which was in Arabic, and report if there were any issues.

I was still trying to observe Ramadan during that time, but it was really hard. I'd be fasting at home, and then I'd go to work on the base. The heat in Baghdad was unbearable, upwards of 120 degrees Fahrenheit. We'd ride around all day inside a Bradley, a turret gun personnel carrier, which felt like an oven. Or sometimes we would walk around patrolling neighborhoods for hours on foot with the sun beating down on us so hard my flesh felt like it was on fire. It was miserable. I didn't have any energy, and I would get terrible headaches.

On one of these days, Bashar and I were driving somewhere, and he wanted to stop and get some beer.

"It's Ramadan," I reminded him.

"I'll stop drinking when I'm done drinking," he said with a grin.

I guess he had a point. Things I'd once thought were important suddenly felt trivial.

As a terp, observing Ramadan quickly became impossible for me. The holy month had been transformed since the war; it was no longer about joy and community. Instead, we were living in a fractured country under occupation where the spirit of togetherness had been replaced by fear and uncertainty. People had lost livelihoods or loved ones—or both. All these factors made Ramadan less significant for me from a religious standpoint. And I began to be less observant.

4

ALL IN

After a few months, I had hit my stride as a terp. I was making good money and having fun with my college buddies. And I decided to upgrade my Opel to a 1991 735iL BMW, my dream car. This bimmer was my baby. I went home to show off my car to my dad, although I was a bit concerned about leaving such a nice car parked on the street overnight. After everyone else was in bed, around two o'clock in the morning, I walked into the kitchen and looked out the window toward the house's main entrance, checking on the car. I could see an envelope wedged between the two sides of the black metal gate.

"Oh man, this is not good," I thought.

I went outside, pulled the envelope out, and felt the distinct shape of a 9mm bullet through the paper.

"Okay, that's really not good."

I opened the envelope and found a note, handwritten in Arabic.

We know about all your movements with the Americans. If you don't stop we will kill you.

I woke my mom and dad.

"I just found this outside the door," I told them and handed them the note and bullet.

Mom started freaking out.

"I told you to stop!" she wailed. "What are you going to do? You need to just quit and settle down."

Dad was in shock and said nothing.

I looked at the two of them, sadly.

"Listen," I said. "I'm leaving. I can't stay here anymore. I don't know who these people are, but as long as I'm gone, you'll all be safe. I'm going to pack my things and head out now. It's still curfew, so I can keep an eye out if anyone follows me. If I hit any checkpoints, I'll be fine—I probably know some of the guys there, and I have my badge. Don't worry about me."

Mom was in tears. She didn't want me to leave.

"Hey, I've got a fast car," I said as I hugged her. "I'll be ok. I promise."

I jumped into my bimmer and started to drive out of the neighborhood when I looked into the rearview mirror and saw lights behind me. I floored the gas and gunned it. I snatched my cell phone off the front seat and quickly called a friend who lived about fifteen minutes away at a car repair shop.

"Hey, unlock the gate; I'm coming to you. I need to park my car inside."

I glanced in the rearview mirror again, and I could still see the lights. I made a few quick turns and then floored it again onto the bridge; my tires actually lifted off the pavement. They were still behind me, but I was doing maybe 130 or 140 miles per hour like I was on the Autobahn. I lost them on the straightaway. I was going so fast that I passed my friend's garage and had to back up. It would have made a great BMW ad.

"What's going on?" my friend asked once I'd gotten out of the car.

I explained.

"And you came here??" he exclaimed in horror.

"Nobody's following me. I lost them," I assured him.

He finally gave in, and I ended up staying with him for a few weeks. At night, we didn't just lock the gates—we chained them for extra security.

After a few months, my responsibilities with Civil Affairs shifted significantly. I was no longer focused on humanitarian missions. Instead, I found myself being sent out to staff checkpoints, to join patrols, and to search houses. This meant my work shifted to a twelve hour schedule, alternating between day and night shifts. Our office was moved to a different building, which had three rooms and a bathroom. We used one room as an office, while the other two were furnished with cots for the terps working the nightshift. Night missions typically involved patrols or raids, while day missions focused on checkpoints or door-to-door searches.

After the invasion, the US had assigned Paul Bremer as the temporary governor. He functioned as head of state over Iraq with the intention of rebuilding Iraq free from the Ba'ath party, which some considered similar to the Nazi party. To shape a democratic Iraq, US officials brought back Iraqi exiles who had fled under Saddam Hussein's regime. These exiles, once persecuted by the Ba'ath Party, were now given positions in the new government. However, having lived in exile for decades, they were disconnected from the realities of the country, which had changed significantly under Saddam's rule. These exiles, advising Bremer, often provided skewed advice to serve their personal agendas or vendettas.

On May 23, 2003, Bremer issued an order to disband the entire Iraqi Army, the police, and the Ba'ath Party, effectively putting nearly 400,000 Iraqi soldiers and police officers out of work. Bremer's decision to disband the Iraqi military and police force, primarily due to their Ba'ath Party affiliations, was not only misguided but devastating. In my opinion, this decision pushed many of these individuals to take up arms, sparking the start of the real war.

Certain parts of the city were becoming hotspots for violence and resistance, so we were now often tasked with searching for weapons, explosives, and bombs. Each unit was assigned a specific area to search, and every house was permitted to keep only one firearm for protection. Sniper rifles were strictly prohibited, and we investigated anything that looked suspicious. These rules were implemented by coalition forces to suppress insurgency and help identify potential trouble spots.

It had become increasingly dangerous to venture out into the community. As terps, we didn't have guns, helmets, body armor, or anything to protect ourselves. While we were trusted to help search out insurgents and weapons, we weren't trusted to carry weapons ourselves. As translators from other units began to be killed, we started to press for measures that would keep us safe. We protested to our US teammates.

"We're in the same danger as you. But you have weapons, body armor, and helmets. We need them, too."

Their response was always the same.

"You can't have a weapon. We'll protect you."

I was just a translator; that was my only job. No matter how big the risks I took for the Americans, they wouldn't risk letting me protect myself, and this really pissed me off.

★ ★ ★

By this point, I was exhausted from arguing with my mom about marrying the cousin she had chosen for me. She hadn't just picked a girl—she had gone ahead and made all the arrangements with her family. In Iraq at that time, arranged marriages were standard practice. I tried to reason with her, reminding her that she had been free to choose my father. But she wouldn't budge.

I knew I couldn't go along with it. I had no desire to marry my cousin, and after a particularly intense argument, I realized something had to change. It was time to cut the snake off at the head and take control of my own life.

I drove my car to Bashar's house, told him about the fight I had had with my mom. I asked him to ride with me, because I was about to do something that would take me to the point of no return.

"I'm about to do something really stupid. I need someone to be with me because this is going to be crazy."

He was the kind of friend who said 'yes' instantly. I knew I could count on him.

We drove out to where my aunt and uncle lived, the parents of my proposed bride. It was a forty-five minute drive, and my cell phone kept ringing the whole way there. It was my mother's number, but I didn't pick up. I knew she was still furious from our last fight, and if she found out what I was about to do, things would get even worse.

Until that day, my uncle and I had always had a good relationship. When we reached the apartment building where the family lived, I knocked on the door and asked him to come downstairs to meet me outside. He came down, and my aunt joined us from the grain silo next door where she worked.

"I just wanted to tell you that my mom and your wife are trying to force me to marry your daughter," I said to him. "I see her as a sister, so I'm not sure what to do. But I know I'm not going to marry her."

My uncle, known for his pride and strong opinions, didn't take it well. "What makes you think I'd give my daughter to someone like you?" he exploded.

"Great, we're on the same page," I replied. "Talk to you later."

Without waiting for a response, I turned back to my car, got in, and drove off. My mother started calling again, but I continued to ignore her, turning the ringer of my phone off. I couldn't help but wonder if my aunt had already told her what happened. Later, I found out that my uncle and aunt had a huge fight after I left. But there was no going back now—I had thrown gas on the fire, and all I could do was to watch it burn.

After a few weeks, the ashes had finally settled a little bit, and my mom asked me to drive her to see one of her cousins. These relatives weren't close, but we had spent time with them during family gatherings when I was younger, especially around the holidays.

My mom's cousin had a daughter named Rana, who was nineteen years old, and had fair skin and long, dark hair that cascaded over her shoulders. We used to play together when we were little. I thought maybe Mom could agree to me marrying her.

I raised the subject carefully.

"You are always worried about me marrying someone we don't know," I said.

She knew instantly what I was getting at.

"No," she said. "I know her mother. You don't want to be with someone like that."

That got me questioning my mother's motives. Was she just totally against me having any choice in the matter? I don't think she knew that I'd already started inquiring about Rana. Rana was available, her mother had said. She had been engaged, but the engagement had been broken off. If I wanted to marry her, that would be possible, I'd been told.

★ ★ ★

When I got back to base, CPT Smith asked me to translate as he interviewed candidates for a new local police force. This task turned out to be more difficult than I anticipated. CPT Smith was from the US, not Baghdad, so he didn't realize that one of his candidates was a known criminal. In our tight-knit communities, everyone recognized the people who caused trouble. You heard stories, you knew people, and news traveled fast.

Right before Saddam fell, he released most of the prisoners that were being held. This added a layer of complexity to an already difficult task. Not only were we trying to rebuild a police force, a military, and a new government, but we also had to contend with criminals vying for positions of power. The US personnel, unfamiliar with the local faces and stories, had no idea who these people were. It made my job, and theirs, that much harder.

My task was to interpret, not just translate. So I wasn't simply providing CPT Smith with a word-for-word

English translation of what was being said. I was reading the situation, acting as a cultural advisor. While it wasn't my job to make decisions, I had a personal investment in the outcome because my goal was to help rebuild my country, and I wanted to do it right.

"This guy, who you're about to give a badge and a gun to, is a gangster," I told CPT Smith in English.

"Everybody deserves a second chance, just translate," CPT Smith responded.

I thought to myself, "You don't give a violent criminal a gun and badge. We are building a police force."

But, of course, it wasn't my decision.

The impact of those hires turned out to be pretty significant later on. The new police force ended up being more of a corrupt security system. I'm not saying the entire force was corrupt, but a lot of people were there to collect paychecks and demand bribes rather than serve the community.

5

CAMP FALCON

After spending about nine months at Camp Slayer doing patrols, checkpoints, and a few raids, I had built some very strong relationships not only with the other terps but with the soldiers as well. When we got the news that this unit of soldiers was leaving and a new unit was coming, it hit me really hard. I had developed a trauma bond with most of these guys, and some had also become real friends.

This became my new normal. The US troops came and went, serving on nine- or twelve-month deployment rotations before going back home. But the terps never rotated; this was our home. We never got a break from combat and lived in a constant state of danger.

On this particular day, we waited with anxious anticipation to find out what was next for us terps as a new unit arrived. At first, it was very hectic, and we could tell operations were shifting. We were told that we would all be relocated to a new base called Camp Falcon, which was located in Dora, a neighborhood in southern Baghdad. Before the invasion, Dora had been an area where many Sunni military officers, who were also Ba'ath Party members, had lived. Given this history, I knew Dora would be a challenging place to work. But I had no idea that agreeing to work there would be like having a death wish.

When all of the terps first arrived at Camp Falcon, we had been assigned to a terp house, which was a little building on the northeastern corner of the camp where we all lived together. We were right next to the guard tower. A few weeks into my time at Camp Falcon, command decided we should move into the barracks with the Army units we were working with in order to build stronger relationships.

We weren't sure if that was the real reason we were being split up or if it was because we were way too loud for the tower guard. Either way, two or three terps were assigned to each barrack. Despite being split up, we made an effort to still hang out with each other between missions. The barracks were in rows next to each other, with shade shelters in between them. When we weren't working, we would hang out in the shelters, smoking and catching up with each other. Sometimes, soldiers would also join us.

Ever since we had met riding the shuttle together in college, Khalid had become one of my dearest friends. We confided in each other about everything. He knew all of the drama between me and my mother, and I knew about his girlfriend. One day when we were smoking cigarettes and drinking chai—the hot black tea with a few sugar cubes that's popular in Arab countries—he shared some news.

"I'm getting pretty serious with my girlfriend," Khalid said, a hint of a smile on his lips.

"That's great, man," I replied, genuinely happy for him. "When are you going to see her next?"

He sighed, the smile fading.

"That's the problem. My car is broken, and taking a taxi from the base could attract prying eyes. If anyone sees me, they might identify me as having connections with the US. That would put both her and me at risk."

He leaned in.

"Can I ask you a favor?"

"Sure man, anything for you," I said, exhaling a plume of smoke.

"Do you think you could give me a ride? I'm supposed to meet her in Al-Monsour at a restaurant tonight."

"Of course, dude. Maybe one day things will work out for me, too."

Just as I was about to tell Khalid about Rana and the drama with my mother, our conversation got interrupted because a soldier on guard duty needed me to translate.

"There is a civilian at the gate. I need a terp," he announced.

Civilians would often come to the gate to report suspicious activity. The reality was that the US Army and coalition forces relied heavily on locals for intelligence, which was crucial for their protection.

On one such day, a local informant had approached the gate, claiming he knew the location of some mortar bases. We relayed this information to the unit commander who immediately dispatched a patrol. The informant led me and some US troops out to a rural area outside the city. As we approached, we could see the mortar tubes rising above the field. And once we reached them, we began inspecting them.

"See!" he exclaimed, "just like I told you."

He slammed his hand on the top of one of the tubes.

In that instant, the shell burst out, severing his hand. I knew enough about military hardware by then to understand what had happened. The insurgents had loaded the mortar, but the shell had failed to launch. When they abandoned the site, they didn't clear the tube, leaving it primed. When our informant slapped the tube, the shell seated properly and launched, causing the tragic accident.

★ ★ ★

Since moving to Camp Falcon, missions had picked up, and we were on patrols with US troops almost every day, all day long. Sometimes they were foot patrols through the farmlands of Arab Juboor, other times they were mounted patrols in Humvees or Bradleys. The point of a patrol was to find 'hot spots' of insurgent activity and draw the fighting to us. We would get shot at and then engage the insurgents.

When we were on a mounted patrol, we learned to recognize places where IEDs were buried underground. The team would secure the location and then radio back to command to send an Explosive Ordnance Disposal (EOD) team to disarm the bomb. Sometimes during this process, we would find ourselves engaged in a firefight, usually initiated by insurgents launching an RPG followed by a small firearms attack.

Within a few months, I noticed these firefights were getting more intense and more organized. Mortar attacks were more frequent; we would get hit early in the morning and late at night. I began to feel like Camp Slayer had been the honeymoon phase of being a terp, and now all of a sudden, I was thrust into a really bad marriage.

It was getting more challenging and more confusing to identify who the bad guys were. A huge influx of men were joining the fight, with tribes aiding and sheltering foreign fighters on both sides. These were the early days of Al-Qaeda in Iraq, the Sunni-backed militia whose main volunteers were either criminals or former Iraqi soldiers from the previous regime, now outlawed by Iraq's US interim governor Paul Bremer. On the other front, the Jaysh al-Mahdi (JAM) or Mahdi militia, led by Cleric Sadr, recruited Shias loyal to their ayatollah. The situation was becoming increasingly chaotic and dangerous, making it difficult to know who to trust and where the next threat would come from.

Iraq's wide open borders made it a playground for every bad actor; warlords and Islamic extremists from neighboring countries wanted to defeat the 'infidels' and push their own agenda. Unfortunately, people like me, who just wanted to rebuild the country and live in peace, were the ones that got caught in the middle of the fight and paid the heaviest price.

As the fighting picked up outside the wire—the military phrase for a secured camp boundary—restrictions were also increasing inside the wire. I remember driving back to the base late at night and parking my car in the lot about 200 yards from the gate checkpoint. It was really dark. I put my badge on as I approached the gate.

One of the soldiers sitting on top of a tank at the gate started screaming at me.

"Stop, put your hands up, untuck your shirt and lift it up. Fucking turn around."

I heard the footsteps of a few more soldiers as I turned around. They were aiming their M4s right at me.

"Wait. I am a terp!" I yelled.

"I don't give a shit!" the soldier barked. That was the moment I knew things had changed. I just didn't know how much worse Camp Falcon would get. All I wanted was to have my old unit back.

All the terps were frustrated and exhausted from navigating the unending changes and restrictions. Since arriving at Falcon, security had become a major concern. We encountered so many IEDs along Route Irish. Every time we went outside the wire there was a high probability we weren't coming back.

One afternoon, as I returned to base, I was talking to one of the terps at the gate.

"Did you hear the terp Skinny Ahmed was killed?" he asked.

"What? On mission?" I was stunned.

"No, apparently he was killed when he was home, no details yet."

"It's getting really fucking dangerous," I thought.

This was no longer fun and games. I wanted to know more details, but I couldn't get any more information. Later that day, Bashar didn't show up for work. He was also confirmed dead.

I asked Khalid about it.

"Did he get killed on foot patrol or get hit by an IED?"

"No, he was gunned down at home," Khalid said, his voice filled with a mix of fear and frustration.

"What?! We need to figure this out. This isn't random" I said, my voice tinged with urgency.

"Do you think there is a mole?" Khalid asked.

"I don't know, but if there is, we're going to fucking find him!" I declared.

Almost every day for a month, we got word that another terp was killed while they were on leave. We were literally being picked off one by one. The handful of terps that were still alive started trying to figure out what was happening. It didn't make sense. It was too coordinated for all of these deaths not to be connected. We were being targeted by someone, but by who? There must be a mole in the unit.

One evening, as the sun set, casting long shadows over the base, Khalid, Najar, and I gathered in my room.

"Someone must be tipping them off," Khalid suggested, his voice barely above a whisper.

Najar nodded. "It's possible. We need to be careful about who we trust."

The fear in our eyes mirrored each other's. We were marked, living on borrowed time. The sense of betrayal was suffocating.

We didn't know who the mole was, but we were certain of the motive: we were seen as traitors for collaborating with the occupiers. That had been the message in the latest propaganda videos circulating, showing terrorists brutalizing and executing so-called traitors. With these horrifying images everywhere, along with the deaths of nineteen translators and the harsh reality that we had no way to protect ourselves, it felt like this unit didn't give a damn about us. Under all this pressure, my mental state was deteriorating rapidly. Khalid, Najar, and I knew we needed an exit strategy. If we stayed at Camp Falcon, we feared we would meet the same fate as our friends—death.

The insurgents' threats didn't just endanger translators and their families but also anyone collaborating with coalition forces and those even remotely associated with Saddam's previous regime. No one was exempt from the danger. I heard chilling accounts of people being abducted for ransom, for revenge, or for just brutal torture.

Day by day, the situation deteriorated further. The urgency to leave intensified.

★ ★ ★

My mother called me, panicked. "They've killed him!" she screamed.

"Who?" I asked.

"Your grandfather," she blurted out through sobs.

I jumped into my car and headed straight to his house. When I arrived, the family had already taken him to the mosque to be bathed and wrapped for burial. They were on their way to the cemetery to bury him. Traditionally, during the three days after a death and burial, the family hosts and feeds anyone who wants to come pay their respects.

During these three days of mourning, I started to ask what had happened to my grandfather. How had he died, and who killed him? My grandfather's neighbor told me he had seen a car pull up in front of his shop where he sold construction supplies, including plumbing and electrical parts. One man was in the driver seat, and two armed men jumped out and went inside the shop. They tried to drag my grandfather out and force him into the car. I assume they wanted to kidnap him, torture him, and then ask for ransom.

However, my grandfather had put up a fight and started to create a commotion. Neighbors started to notice. The man driving the car yelled at the two gunmen.

"Just put a bullet in his head!"

And one of the gunmen shot my grandfather at point blank range. The two men then scrambled into the car and peeled off, leaving my grandfather lying dead in a pool of his own blood in front of his shop.

I suspected the JAM because this was their usual tactic: kidnapping or killing people. My grandfather would have been a target given his diplomatic position under Saddam, and he had been an active member of the Ba'ath party until the '90s.

My grandfather's death was my family's first loss in the war; by this point in the war most everyone had lost at least one family member. Many had lost more than one. I was sad, but more than that, I was angry.

★ ★ ★

After the death of my grandfather and several of my fellow terps, life started to feel really chaotic and out of control. One way I decided to take back control was to get engaged to Rana, my distant relative. I hoped that would give me a sense of stability and end the constant nagging of my mother.

Normally, when you get engaged in my culture, it's a family affair. The suitor, accompanied by his parents, visits the woman's home to meet with her family. During this visit, he demonstrates his suitability as a potential husband. If both families agree to the marriage, the fathers shake hands, and a verse from the Qur'an, the Fatiha, is recited to signify the engagement. This is followed by a celebration with juice and desserts. As part of the tradition, the man 'marks' his fiancée by bestowing a gift upon her, perhaps giving her a gold necklace or a ring.

My engagement was nothing like that. Given the hectic circumstances and my limited leave schedule, her father agreed that I could marry her and the next step would be going to the courthouse.

The closer our proposed wedding day came, the more I wanted to make things right with my parents. So one evening, I drove my bride-to-be and her mother to our home. My mother and father were helping my sister with her homework when we walked into the house unannounced.

"You brought them to my house?" My mother was clearly vexed.

My fiancée's mother tried to calm her down.

"Please," she said, "let's celebrate their engagement."

"No," my mother cried. "That will never happen. Get out of my house!"

In that moment, I lost all hope of reconciliation and told Rana and her mother that we needed to leave. I drove them home, then drove back to the base in silence.

The next day, some of my soldier friends noticed I was pretty quiet.

"You don't look okay," they said with concern in their voice.

I gave them a quick run down of the situation: how my parents had arranged a marriage for me; how I had broken it off and made my own choice; and how my parents were now furious and wouldn't accept the girl I had chosen.

"That's not right. You should marry whoever you want," said an older American woman who worked as a contractor on base.

Maybe so. But that was how things worked in Iraq. I thanked her for her support and walked towards the shelter. I sat in the shade, as I lit one cigarette after another. I weighed my options: marry my cousin and be trapped for the rest of my life or marry someone who seemed nice enough and to whom I was actually attracted. The choice became clear. My freedom to choose was more important than my parents' blessings. I had never been able to satisfy my mom, and nothing I did ever seemed good enough for my dad. It was time to buck cultural norms and do it my way, even if it meant losing my family. My mind was made up. There was no turning back.

★ ★ ★

Back at Camp Falcon, Najar, Khalid and I weren't the only ones who'd gotten nervous; even the new terps were scared shitless. We decided it was time to stop working as interpreters until we could figure out what was happening. Why were so many of our friends being killed?

Najar and I went to talk to our commander. We had decided to turn in our ID badges.

"We're getting killed left and right," we told him. "We can't continue doing this work until whatever's going on gets sorted out."

The commander was not pleased. He needed us terps for his mission to succeed. Interpreters were critical because we didn't just translate words but cultures, histories, and intentions. We often found ourselves literally caught in the middle, working to protect innocent Iraqis while helping the soldiers identify the enemy.

Knowing we were going to walk, our commander started to threaten us.

"If you leave, I will red flag you!" he told us.

We knew that would be a career-ender. If we got red flagged, we could never work for coalition forces or step foot on a coalition base again.

"Why does that even matter?" I barked. "If I stop working for you, I won't have a job. But if I keep going, I'm dead. The terrorists are trying to take our lives, and now you are trying to take our livelihood. And you don't give a shit about us. I'm done!"

And with that, Najar and I slammed our ID badges on the table and walked out. We didn't know if he had just been bluffing or if we would really be red flagged, but we knew the only way to survive was to get out, lay low for a bit, and hope they would find the mole.

PART 2

Controlling Destiny

(2004 – 2009)

6

ON THE RUN

We headed to Najar's house mainly for practical reasons. Returning to my parents' house was out of the question after the threatening letter and the bullet—and of course the encounter with Rana. My mom and I were still fighting about her. We figured Najar's house would be safer.

As we pulled inside the house gate, his brother Wissam closed it behind us. He offered to stay in the yard for a while as a lookout. We agreed, and he stayed behind, pacing back and forth, keeping an eye out for anything that might look suspicious.

Thankfully, Najar's parents welcomed us into their home. His mother, a sweet and always-smiling woman, was warm and hospitable. She kindly offered us some chai.

Within thirty minutes, Wissam came inside to join us.

"There's a car with four armed guys in it; it's driving up and down our street. They keep looking at Najar's car," he said.

"Fuck," I replied. "We were followed. We gotta go. It's better for us to be together and away from them."

We immediately grabbed all our stuff and waited by the door for the right moment to leave. Wissam monitored the street as the car that had followed us continued to circle the block. He saw it take a turn at the end of the street.

"Now!" he said.

We bolted out of the house, jumped into the car, and took off in the opposite direction, driving like a bat out of hell. Next time they looped past the house, we weren't there.

Najar and I agreed that it was too dangerous to hide out at our relatives' houses, so we agreed to try couch surfing across Baghdad. We would call different friends and ask to stay with them for a few days. Most of them agreed. One friend, a contractor at Camp Anaconda, even let us crash in his office in Zayuna. What was supposed to be just a few days turned into three weeks. We were constantly on the move, always looking over our shoulders.

As far as I knew Khalid, Najar, Ethar, Sa'ad, and I were the only translators from our initial group at Camp Falcon who were still alive at that point. The other nineteen translators had been killed.

While Najar and I continued to find random places to crash, Khalid went home and lived with his family. He was preparing to propose to his girlfriend. I wanted to get married, too, but now I was living like a fucking fugitive. I desperately wanted to find a place to rent in Ghazaliya, a suburb of Bagdad, close to Rana's parents, so I could marry her. In a matter of a month, I had gone from being a good prospect for Rana, making $40 a day, to a jobless, homeless man on the run.

Najar and I were barely scraping by on our savings, and it was tough. Money was tight, and most days we just hung out in the rental house that was near the dangerous area of Al-Shu'la. Khalid knew about our situation and was incredibly kind. He visited us regularly, bringing groceries and cigarettes. If we weren't home, he'd leave the bags at the kitchen door, just inside the front gate. Some evenings, he even stayed to join us for dinner.

On this particular day, Khalid was busy running errands in preparation for proposing to his girlfriend, and Najar was out, too. I stayed home, with neither the money nor the desire to go out. We had planned to meet at the house for dinner, so I spent the entire day chain smoking and listening to music while waiting for my friends to arrive.

As the evening wore on, I started calling Khalid and Najar. Khalid didn't answer, and when Najar finally picked up, he seemed rushed.

"I'm going to be late," he said shortly.

After that, I couldn't reach him again; it was like he had disappeared. For the next hour, I kept trying to call him, but there was no response. Two hours passed, and panic started to set in. I knew something was wrong. Had they been captured? Should I leave or stay put? My anxiety skyrocketed. I felt completely exposed, stranded in a neighborhood where I knew no one and had no way of protecting myself. I was a sitting duck.

Finally late into the night, I received a call from Najar.

"Open the front gate," he said urgently. "I'm going to pull into the driveway, and then we'll talk. Hurry!"

"Is somebody chasing you?" I asked, panic rising. "Don't bring anyone here!"

Moments later, Najar drove through the double gate I had left open and pulled up to the kitchen door. He stepped out of the car, leaving the car door and the gate open. He leaned over the car's roof, resting his elbows on it as he buried his head in his arms. He seemed despondent.

"Dude, what's going on?" I asked, anxiety mounting.

"They just killed Khalid," he replied.

I was stunned.

"What? Let's go inside," I said, not wanting the neighbors to overhear us.

Shortly afterwards, Khalid's brother called me. He explained that Khalid had gone out to buy dinner in the Ameria neighborhood, only a few kilometers from our rental house. As Khalid walked down the main street, trying to decide between the restaurants, he sensed someone was following him. He ducked into a restaurant to throw them off, but then quickly left and ran into the neighborhood's maze of streets, trying to shake his pursuers. But they chased him down, and as he attempted to climb over a wall, they opened fire and killed him.

Now Najar and I were alone in the house, staring at each other in shock. I began to tear up remembering our last mission. That particular day, the mission hadn't been that big. We just needed to raid four or five houses to flush out any potential insurgents, and it didn't take all that long. When we were walking back to the parked Bradleys afterward, Khalid had asked me for a cigarette.

"I left mine in my jacket," he'd explained.

Khalid had started off the day wearing his brown leather jacket—one of those nice things that our well-paid jobs allowed us to afford.

"Where is your jacket?" I asked.

"There was a lady outside the house, and she was cold," he'd said. "I put it over her."

Khalid was genuinely a good guy. Like all of us, he believed that our work as translators would deeply benefit our people and our country.

I snapped back to the reality of the dark house.

"Najar, we need to leave. It's not safe here anymore."

He nodded but slumped into a chair. We were both overwhelmed with grief and couldn't think clearly. We decided to figure it out in the morning and try to get some sleep.

The next morning, we woke up to find Khalid's girlfriend at the door. She came inside and started to unravel.

"What's going on?" she demanded. "Where is he? Did he really die?"

"Yes, I'm sorry," I told her. "That's what his brother told us."

She broke down, crying uncontrollably as she paced through the house. She and Khalid had shared private moments here, and she was desperately searching for any of his belongings he might have left behind. When she finally left, Najar and I knew it was time to leave. The place was no longer safe. We didn't know who had gotten to Khalid, but it was likely the house had been compromised.

That day, Najar and I returned to my friend's office in Zayuna. He told us we could stay as long as we needed. It wasn't home, but it was probably safe.

I fell into a troubled sleep on one of the office couches. Khalid came to me in my dream, speaking a mix of Arabic and English slang. He was clean, smelled like rose water, and dressed in all white.

"Dude, I'm leaving," he said. "Do you need anything before I go?"

I woke up screaming his name, my entire body vibrating with anguish and terror. The room was empty except for me and a guard asleep on the other couch. I couldn't stop crying, gasping for air as grief hit me like a ton of bricks. I stumbled out into the open street, desperate for air, and collapsed, sobbing so hard I threw up. My entire body was trembling and in that moment rage enveloped me like a dark, comforting blanket. All I could think about was avenging my friend's death. Killing the bastards who murdered Khalid—this was all that mattered now.

A few days later, Najar and I started trying to figure out who killed Khalid. Who was the mole that was clearly feeding the terrorists information about us? The only two translators from our group that were left at Camp Falcon, as far as we knew, were Ethar and Sa'ad. So It had to be one of them. By process of elimination, we would find out which of the two it was.

7

A MOLE AMONG US

Sa'ad and Ethar had remained close friends while working at Camp Falcon, so it wasn't surprising that Sa'ad often sought out Ethar and invited him to parties. Ethar wasn't the type to go partying, even with the promise of girls and booze. At that time, Ethar never left the base, not even for leave. He had a critical role as the translator for the Command Sergeant Major, who valued Ethar's work and loyalty. There were even rumors about efforts to secure Ethar a visa to the US. Because of this, Ethar was cautious and avoided any potential dangers in the local community off base.

Sa'ad, on the other hand, seemed to be off base constantly. Some of his absences could have been due to missions, but it felt like he was away more often than our jobs required. We knew Sa'ad had been urging Ethar to join him at parties for a while.

"You should come," he'd say, promising wild, crazy parties that would have enticed most other guys.

"We should go," he'd push.

One day, for reasons unknown, Ethar finally caved.

"Alright. Fine. I'll go," he said, leaving his car in the parking lot just outside the checkpoint and heading off base with Sa'ad.

But it wasn't the kind of party Ethar expected.

Ethar didn't return to the base that night. And his mother and two sisters began calling Najar and me, asking if we'd heard from him. We were still on the run, and it had only been a few days since Khalid had been murdered.

"I haven't heard from Ethar," I told them.

I tried calling his phone, but there was no answer. With Ethar missing, a sinking suspicion began to form—and I suddenly couldn't shake the feeling that Sa'ad might be the mole.

Najar and I called the translator who we knew worked the main gate at Camp Falcon to find out more information.

"The whole base is going crazy right now," he said. "We are all on alert. The Command Sergeant Major is pissed that it was his terp. Good thing you guys aren't here."

We eventually learned that Sa'ad had gotten into a heated discussion with some of the new terps who shared a room with Ethar. While Ethar was showering and changing, Sa'ad had sworn the others to silence.

"Keep this secret," he said. "Don't tell anyone we're going to this party."

"Why? Why does it matter?" the other terp asked. "Who cares?"

"I'll kill you if you say anything," Sa'ad warned. "Keep it quiet, man."

Of course, the terps should have said something, recognizing that Sa'ad's behavior was off, but instead Ethar went blindly with Sa'ad, his so-called friend.

Security cameras showed Sa'ad and Ethar leaving base in Sa'ad's car. During the investigation that followed, it became clear that Sa'ad was the last person Ethar had been seen with. The soldiers decided to raid his house, and during the raid, Sa'ad tried to play dumb.

"I'm a translator!" he screamed. "I work for you!"

"Yeah, we know who the fuck you are! You're the person we've come for!"

One of the soldiers hit him right between the eyes, and his nose started bleeding.

"Where is Ethar?" they demanded.

"I don't know."

As they questioned him, Sa'ad finally admitted he'd been working with insurgents and that he had delivered Ethar to them to be tortured and killed.

"But I only did it because they had threatened to kill me and my entire family. They would do, too, because they already kidnapped my brother," he insisted.

This is how the terrorists worked: threats and coercion were used to make civilians do the terrorists' dirty work. Sa'ad faced a difficult choice. Ultimately, he chose to give the terrorists all of the information about the terps: our travel schedules, our home addresses, our vehicle descriptions and our family details. This is how they managed to kill twenty-one terps in two months outside the wire. We were literally being hunted like animals.

But Sa'ad wasn't a terrorist; he was operating under threat from the terrorists. I just didn't understand why he didn't tell us what was happening. The US Army could have protected him and his family. Instead he let all of those terps, my brothers, get murdered in cold blood.

That's the last we ever heard of Sa'ad. I think they sent him to Camp Bucca, which was a US military prison in Bassra, about 350 miles southeast of Baghdad, near the border with Kuwait.

8

CAMP JUSTICE

Najar and I were fed up with living on the run and in constant fear. Now it was just the two of us, and it sucked. We made a bold decision to drive to the Green Zone where Titan was stationed. We wanted to rejoin a unit focused on conducting raids. We just hoped that we weren't actually red flagged. All we wanted was a chance to fight back— we were done being hunted.

"If we die we die," we thought. "But we wanted to at least die fighting." We were prepared to face the risks head-on.

On the way, Najar looked over at me.

"It's just you and me, Waleed. This is it. We are either going to survive, or we're going to die together. Either way we're gonna kill some sons of bitches in the process."

I looked at him, smiling.

"Alright, buddy, I'm with you. Let's do it."

We were determined to go out with guns blazing and avenge the death, not only of Khalid, but of the entire group of twenty-one translators from Camp Falcon.

We arrived at Titan and told one of the men there our story. And that we wanted to be terps again. We were done with patrols; we wanted to be put with units that were doing raids and kicking in doors. We'd been hunted, and now we wanted to do the hunting.

Neil, the office manager, gave us a language proficiency test to complete. He said we'd talk after. Najar and I both turned in our test and anxiously waited to find out whether we would receive assignments. We just assumed we would be together, since we'd been the only ones of the original group of terps to survive Camp Falcon.

Neil motioned for me to come towards him.

"I've got a spot for you," he said.

I nodded. "Okay. What about Najar?"

"We've got a spot for him, too, but it's at a different place."

My heart sank.

"What, the fuck" I thought. I didn't want to go anywhere without Najar, but I had no choice.

"Okay," was all I managed to say out loud. That was the last time I ever saw Najar. To this day, I have no idea if he lived or died.

Neil told me I'd be working with both the Iraqi Special Operations Forces (ISOF) and the American Special Forces (SF).

"Go to the base located in the Shia neighborhood," he explained.

I assumed he meant the base in Sadr City, so I got into a cab and made up a story about my brother being captured by the Americans and needing to go to the base to get him.

The taxi driver, sympathetic, started calling the Americans 'pigs.' Sadr City had been a pretty dangerous place for a long time. It was a low-income, mostly Shia, suburban neighborhood called Al-Thawra, which meant 'The Revolution.' During Saddam's regime, he'd named it Saddam City in his own honor. That was his way of showing his spite and control over this Shia neighborhood.

But when I got all the way to the base, I found out that I had been mistaken. Now, here I was, stranded outside of a random gate at a regular Iraqi army base, feeling exposed and vulnerable. I approached the gate guard who looked at me with confusion. He, of course, had no idea who I was.

I quickly turned around and called Neil

"Hey, this is not the right base," I said once he picked up. "You've got to help me. Do I head back?"

"I'm sorry for the mix up," he said. "It's the one near Kadhimiya close to another Shia area. I think it's called Camp Justice."

That was all he said.

"Shit!"

I was so freaking frustrated. I had just risked my life to go to the wrong base.

I walked back to the same taxi and got in, relieved that it was still waiting. I told the driver there had been a mix up, that my brother was in Kadhimiya. He nodded. Sitting in the back seat, my mind was racing and I hoped the driver was not getting suspicious.

We drove back through Baghdad, heading toward Kadhimiya. Our destination was Camp Justice, located just past the Al-Aimmah Bridge. It was a place that had terrified every Iraqi since Saddam came to power because of the rumors we'd heard about its secret jails and torture chambers. People even said there was an oversized meat grinder, designed to dispose of the corpses of Saddam's political opponents. It still felt eerie.

After crossing the bridge, I knew the base would be on the right. As we approached a traffic circle, my nerves started to get the better of me. I asked the taxi driver to stop the car.

"Just drop me off here," I said, my voice tense.

I decided to get out at the traffic circle and walk the rest of the way. I knew this area of town pretty well because it was close to where some of my relatives lived.

I finally made it to the main gate which was near a public park. I saw an Iraqi soldier standing outside wearing a Lithuanian uniform, which was odd to me. I could tell this was not a regular Iraqi Army base, but I had no idea this soldier was part of the Iraqi Special Operations Forces (ISOF). At first, I was hesitant to talk to him, but finally, I approached.

"I have an appointment with someone called Chief Brad."

"What's your name?"

"Waleed."

"Okay, hang on," he said, as he called it in on the radio.

As I waited at the gate, I noticed a guy coming towards me dressed in shorts, flip-flops, and a T-shirt, carrying just a sidearm. He was seriously jacked. He looked like freaking Arnold Schwarzenegger.

"Holy shit, who are these people?" I thought.

Before this, I'd only seen regular soldiers dressed in full gear. This was my first introduction to the guys who worked for the US Special Forces.

He shook my hand. "Welcome to Camp Justice," he greeted me. "I am Jerry. I'll bring you to the chief. We walked toward a white Land Rover Defender and got in. Three minutes into our ride, we passed the Iraqi SF area. We drove further into the base's interior where the American Special Forces unit was housed. I saw a courtyard with a burn pit on the right. In the middle was an open area, surrounded by a three-building structure housing the SF Team, the Operation Detachment Alpha (ODA) barracks, the terp house, and the Puerto Rican guard house. On base there was also an ODA operations office, a commando chow hall, gun ranges, a gym and a parking lot.

I got out of the car and followed Jerry through another little gate. I looked to my right and saw a body opponent punching bag in what looked like a little makeshift outdoor gym.

I walked a few more feet towards a building with a camouflage tarp on top of it. I then went through a little hallway and saw a fair skinned man with blonde hair and blue eyes, whom I assumed was Chief Brad. He shook my hand.

"I'm Chief Brad, your POC." He pointed to the small table and motioned to sit.

He leaned in close. "So, tell me your story. Why are you here?"

I looked him square in the eye.

"I'm a translator, and I need a job."

He cleared his throat. "Now, again without the bullshit."

A little confused, I took a deep breath and continued.

"Well, I worked with the regular army and a lot of my friends were killed. I got tired of it. And I don't want to do pointless checkpoints. I want to do raids and go after terrorists." I said.

Chief Brad nodded with a slight smile.

"Okay. I can understand that. Have you ever handled a weapon?"

"No," I explained. With the regular army, they did not let us have weapons.

"It's different here. You will be expected to be part of the team." Chief Brad moved closer and continued, "We are advisors, and our job at this base is very specific: to build and train an Iraqi commando unit. This camp is divided into two sections. Half of the facilities are for us—the Special Forces and the Iraqi SF commandos—and the other half of the base houses a regular US Army unit."

Chief Brad told me to go home, get my hygiene kit and pack a bag for a few weeks' stay on base, and then come back.

"I can't go home." I told him. "I don't have a home anymore."

He didn't seem fazed. "Go get what you need and come back," he insisted.

I shook his hand. "I'll be back," I said.

★★★

I had nowhere to go except to the house where my fiancée still lived with her parents. I showed up there feeling a mix of desperation and shame.

"Could I crash on your couch for a few days until I head back to the base?" I asked.

They were understanding enough, but the situation was undeniably uncomfortable. Here I was, the guy who had proposed when he had a good-paying job with promising potential. Now in a rough patch, I was more of a liability to their family.

I stepped into the living room; the scent of cigarette smoke heavy in the air. My soon to be mother-in-law sat on the couch, eyes glued to the flickering TV screen. The same woman who had been so joyful and excited at our engagement now seemed distant and detached.

"Thank you for letting me stay." I said softly, trying to break the silence.

She took a long drag from her cigarette and exhaled slowly, her gaze never leaving the screen. Silence. I stood there for a moment, feeling the awkwardness settle over me like a thick fog. I couldn't shake the feeling that I was an unwelcome guest.

Her family was aware of the threat against me, but they lived in a different neighborhood where I was not known, so I didn't think they were worried about their own safety. Honestly, during that time, everyone in Iraq faced danger. If you were wealthy, you faced the constant threat of being abducted for ransom. If you were poor, you risked encountering an IED that could explode, killing you and everyone nearby. Even something as simple as walking or driving through a neighborhood with an insurgent checkpoint could lead to disaster.

In some respects, being on base offered a sense of security compared to the unpredictable dangers outside. But here, in this house, surrounded by the silent disapproval of my fiancée's family, I felt more exposed and vulnerable than ever.

Rana came in. "When will you start your new job and how much will you get paid?" she asked.

"Not soon enough," I thought.

I just wanted to get back on base and sort out my new position with the Special Forces so I could prove to Rana's family I was worthy of marrying her.

After a few days, I arrived on base with my bag and hygiene kit. I was immediately taken to the S4 warehouse to be issued my uniform, boots, vest, and helmet. Then they took me to the armory where they assigned and registered a weapon to me.

Special Forces are distinct from the regular Army. If the Army is like antibiotics, Special Forces are the scalpel that removes the cancerous tumor. I couldn't be a terp with this elite unit without knowing how they operate nor without knowing how to use a gun.

Prior to Camp Justice, I was just a civilian contractor who worked for the military. I wasn't trained to be a fully contributing member of a military unit. My job had just been to follow the team leader and to translate for them. They would handle my protection. I'd be in the firefight, but instead of fighting, I'd be ducking, hoping the other soldiers were covering me. In battle, I had been a liability.

Here, I was expected to function not just as a translator but as a teammate. I was to become a military asset to the US and Iraqi commandos. It became clear from day one, that when I was outside the wire, I'd carry a gun like the other soldiers. If the unit was in a fight, everyone, including the terps, was going to shoot back. The ODA treated me like one of their own.

Their message to me was clear. "Sit down. Get to know us. And if you go out with us, you're part of the team."

The first few weeks at Camp Justice were incredibly lonely because I didn't know any of the other translators. I missed Najar and Khalid terribly. After about a month, though, I finally connected with three new terps: Mazar, Daryan, and Fahad. Mazar and Daryan hailed from Kurdistan, while Fahad was from Baghdad. All three had extensive experience working with Special Forces and had been on numerous missions. They were accustomed to being fully integrated with the team and had earned their trust and respect.

For the first month, I shadowed Mazar for training. I immersed myself in debriefs, attended meetings, learned operational procedures, and familiarized myself with the training schedule. The regular duty rosters for the battalion were pivotal in determining assignments. Each month, the schedule was divided into four segments for the battalion companies: missions, training, Quick Response Force (QRF), and leave.

Missions were at the core of what we did, ranging from high-stakes raids to multi-day operations that took us across the country to engage in larger-scale activities. Training sessions were dedicated to skill refinement and to preparation for future operations, offering a crucial opportunity for growth. However, the most intense assignment was QRF, where we had to be ready to respond to an emergency at a moment's notice, dropping everything and rushing to wherever we were needed most. After that kind of constant, high-stakes pressure, leave provided a much-needed respite, offering a break from the relentless demands of being a translator.

I began to learn proper military acronyms and specialized SF keywords that I needed to be able to translate, like SSE, which stood for 'Sensitive Site Exploitation,' and CCP, which meant 'Casualty Collection Point.' I was trained to be a valuable asset in battle, learning protocols for various combat situations, as well as how to handle firearms effectively. On top of everything, I translated this training for the Iraqi SF units.

Every Iraqi SF company had its own arms room near the barracks. We would go here to get our weapons and ammo, then be transported to the range for training. After training, we would clean our weapons and return them to the armorer, who would lock them back up.

My first mission with the Special Forces was a quick air assault. While I don't recall the specifics of the operation, the flight itself was unforgettable. It was my first time in an aircraft, and when they said we'd be flying in a Blackhawk helicopter, I was terrified. My seat was in the back middle, and the tilt on takeoff felt like a rollercoaster just after cresting a rise. I had never experienced anything like it.

On the way back, I asked to sit near the door so I could see my city from above. I had never flown over Baghdad before, and the view was breathtaking. Seeing all the palaces as we approached the airport was truly amazing.

Later on, helicopter flights became routine for me. I no longer stayed in the back of the helicopter. Instead, I sat in the doorway, my feet dangling down, breathing in the adrenaline.

Even though I was excelling with my new team, Chief Brad thought I'd be a good fit for the scouts. The scouts operated differently from the rest of ISOF. They trained and operated covertly. The scouts wore civilian clothing and carried concealed—pistols or the short versions of the AK. They learned to drive civilian cars at high speeds for things like chases or 'snatch and grabs.'

I was hesitant at first; I didn't fully understand what the scouts did. I didn't want to give up the adrenaline rush of conducting raids or the satisfaction of dealing directly with terrorists. Chief Brad assured me I'd still be in the action and that the scouts would give me a chance to learn some new skills and reconnaissance tactics. So, I agreed.

When I first joined the team, the scouts were assigned recon missions. But eventually we transitioned to collecting intel. We would watch possible targets and take turns surveilling their houses.

On one such mission, we had to retrieve one of our scouts who'd been captured and locked up in the local police station. He was Kurdish, and his accented Arabic had made him stand out. While surveilling their neighborhood, some locals had found him suspicious and reported him to the police.

The Iraqi police detained and interrogated him but eventually allowed him to make a phone call. We were ready to conduct a hostage rescue, if necessary, but we managed to verify his identity and confirm his story, securing his release peacefully.

In Iraq, there was no such thing as minding your own business. Everyone was involved in everyone else's affairs. If you were not from the area, you'd stick out like a sore thumb.

I stayed with the scouts for a few more months until Chief Brad assigned me to Delta Company. This led to my first large-scale operation that I participated in with Special Forces, which took place in a town near Ramadi. Intelligence indicated that fighters were crossing the border, bringing weapons into Iraq. Lawlessness was on the rise, insurgents were growing bolder, and the police were under constant attack. The Marines had already locked the city down—no one was entering or exiting.

Our forward operating base, or FOB, was near Rutba, which was called the 'Korean Village,' close to the borders with Syria, Jordan, and Saudi Arabia. We arrived in the middle of the night to find the town in chaos. The police station had been overrun. A police car was on fire, and the station's windows were shattered, its doors wide open. The city was clearly under insurgent control.

As our Humvees and trucks pulled up, I translated for the Delta Company Commander, helping organize the teams. We had several high-priority targets to hit immediately—houses identified by Marine intelligence as harboring insurgents.

After raiding the first house, we suddenly came under fire. We quickly moved from one house to the next, searching for weapons and insurgents. Every step was a calculated risk as we engaged in intense firefights, clearing rooms and making captures. The sound of gunfire echoed through the night, each moment a blur of adrenaline and chaos.

By the time dawn broke, we were all exhausted, running on fumes. I caught sight of Daryan, one of my terp friends, his weapon held in the 'low-ready' position as he leaned against a wall—fast asleep. The scene was too surreal to ignore, so I snapped a picture—an image of a soldier too tired to stay awake, even with danger all around.

This relentless pace continued for three nights. Each night, we raided houses and engaged in door-to-door combat, detaining anyone we positively identified as a terrorist and passing them to the Marines for transport.

By day, we retreated to our convoys outside of the town and lay down under the trucks to sleep, seeking shade from the desert sun. The heat was merciless.

On the final day, it was already daylight when we finished our raids. Instead of going to the trucks, we stayed in one of the large houses we had cleared. The cool tile floors and walls were a welcome relief. We confined the family to one room, while the soldiers spread out in the halls and other rooms.

Remarkably, the husband of the family was kind enough to make us chai. That's Iraqi hospitality for you—a group of armed soldiers invades your home, and after they recognize you're not an enemy, you offer them tea.

9

WEDDING BLISS

Every month we got one week off, so I'd planned to get married during my next regular leave cycle. Without my parents' approval, the wedding was all up to me. I didn't care about having a traditional Iraqi Islamic ceremony. Culturally, most people would have an imam presiding over the ceremony but I was a rebel. So when the day came, we went to the courthouse. It was nothing special, but it was official.

A few weeks later, we hosted a celebration at a hall I had rented. It was a party with music, dancing, and food. A bunch of my college friends showed up, but none of my family came. My new mother-in-law's behavior that day was bizarre. She wasn't happy; she was snappy and dismissive.

There was not a lot of joy during the celebration; it was more like going through the motions. I did love Rana, but my motives weren't solely based on love. I was also a man looking for an out from marrying my mom's preferred cousin. Rana was my solution.

Our marriage seemed doomed from the start. Even our wedding night was a disaster. We arrived at our hotel, and Rana kept saying, she felt sick and had a headache. Nothing happened. We didn't consummate the marriage for almost two years. Granted, I was only home for five months out of those two years due to my rotations, but it always nagged at me—was she perhaps still in love with the man she had previously been engaged to?

Right after the wedding, I had to report back to the base, so there was no honeymoon, no time to build a foundation. Once I returned to work, everything with Rana changed. As we talked on the phone, I noticed her true personality started to show. It felt like I wasn't married to just Rana—but also to her mother, who seemed to control every decision Rana made.

I had furnished our new rental home, but one day I returned from the base to find that half of our furniture had been replaced because Rana's mother wanted something different.

"This is our house; not your mother's house," I told her.

It would've been different if I'd known in advance that she'd wanted new furniture. But coming home to it without warning was infuriating. Her mother controlled her every decision, and she tried to control me, too.

Rana spent every spare moment at her childhood home, a large house divided into multiple levels down the street from where we lived. Her family's living space was on the top floor. She'd go there nearly every day, all day. Even when I was on leave.

"I don't want to spend what little time I have at home with your parents," I told her plainly. "We have our own house. I just want to come home and relax and be together."

She disagreed. So when I was home, she would still go over to her parents' house and leave me all alone. I hated it.

Eventually, I realized that Rana was still just a child, pretending to play house. My mother had been right—I'd made a terrible choice, and my marriage was unraveling fast. Weeks went by without a call from her, and I wasn't going to be the one to make the first move. I had made my choice, and now I was going to live with it. I didn't want anyone, especially my mom, to know the truth because that would mean admitting defeat.

After about a month, my mom finally broke the ice. I don't know who told my mom that I'd gotten married, but word traveled fast. Maybe a distant relative let her know.

She was still devastated that I had gotten married and was crying on the phone.

"How is work? Are you safe?" she asked.

I assured her I was fine.

"I've accepted that I can't change your decision," she continued. "But don't expect me to talk to your wife."

I was just relieved that she couldn't bring up that marrying-my-cousin bullshit anymore.

It took a few months for my mom and I to really mend our relationship. I kept sending them money to help support the family, knowing my dad was deep in debt and they were struggling. My marriage hadn't affected my relationship with my dad very much. He was distant before the wedding, and he stayed distant after. He was too consumed by his own world to care.

On the other hand, my father-in-law wanted to connect with me because he aspired to be a translator. He was looking for a way to support his family. I hadn't even realized he spoke English, but apparently he did. I explained the process to him. It had changed since I'd started; now, you had to go to the Green Zone for an interview before getting assigned to a unit. Unsurprisingly, he was assigned to Camp Falcon, known for its high turnover of translators.

Later that day, I went back to base, and one of the terps shared some news.

"Did you hear we are moving?" he asked.

"Oh please, not Camp Falcon again," I thought.

"We are moving to Baghdad International Airport Area IV," he said.

"Why?"

"It was the Iraqi Ministry of Defense's decision," he said. "They built a huge base for us, better equipped for training. It's also next to the ICTF, the Iraqi Counter Terrorism Force."

For a moment, I felt relief, knowing I wasn't being sent back to Camp Falcon.

★★★

That night, around midnight, Rana called me, screaming hysterically.

"What is going on?" I interjected between the sobs.

She didn't answer. I heard a bunch of commotion in the background and then the phone hung up.

I tried calling back, but there was no answer. My first thought was that she was being kidnapped. I needed to check on her, but it was late, and the entire city was on curfew by now. No one could go anywhere after dark, and I was not supposed to leave the base.

I thought, "Fuck it. She's my wife. Our marriage might not be perfect, but I'm loyal. If she's in danger, I'm going to save her, no matter the risk."

So, with just my pistol and ID, I left the base on foot. After a few blocks, I spotted an Iraqi police checkpoint ahead.

"Shit," I muttered under my breath, knowing that some police officers were aligned with militias or had criminal backgrounds. It was impossible to know if they were trustworthy or not.

My heart raced as they stopped me and questioned why I was out past curfew. I quickly explained that I believed my wife was in danger, showing them the fake Iraqi lieutenant ID I often used as a cover. To my relief, they nodded understandingly and agreed to drive me to my in-law's house.

As soon as we got to the house, the police insisted on escorting me up to the second floor. I opened the door and heard screaming and yelling, so I took a few more steps inside with the police behind me.

Rana came around the corner.

"What the hell are you doing here with the police?!" she cried. "Get them out of the house!"

"I thought you were being kidnapped, and they were kind enough to drive me here."

I quickly realized I had been mistaken. It was a domestic dispute between her and her mother. And I turned towards the police officers, embarrassed.

"Thank you so much for giving me a ride. It looks like it must have been a misunderstanding," I said as I walked them to the door to escort them out.

What the hell had I walked into? To this day I still have no idea what they were fighting about but I had never seen rage on Rana's face until that moment. Adrenaline surged through my body, and anger consumed my mind. I couldn't believe I had left the base and risked my life for this. And somehow, in the middle of it all, the fighting turned towards me for showing up with the police, when really I had just been trying to rescue her. Another telling sign that wedding bliss was not in our future.

10

THE DIRTY BRIGADE

We moved from Camp Justice to BIAP Area IV inside the Baghdad International Airport in September of 2005. The plan was to move the whole battalion in three trips. I was in the last convoy to leave Camp Justice. As we rode in the convoy, we passed Camp Slayer, my first base, and then, upon arriving at BIAP, we passed through Gate 1, which was teeming with activity, and drove about twenty minutes inside BIAP.

Gate 1 was the main entrance to the airport grounds and was located on Route Irish. It was marked by the 'Flying Man' statue, as the troops called it. The statue honored the ninth-century engineer, astronomer, chemist, and poet Abbas Ibn Firnas, who was the first man to attempt unpowered flight.

As usual, there was a lot of hustle and bustle at this main gate the day we arrived. Taxis constantly came and went, picking up and dropping off passengers using the civilian side of the airport. Soldiers and contractors were also queueing at the gate's checkpoint. In order to get all the way back to Area IV as a translator, you had to go through multiple checkpoints; that took between one and three hours to get through, depending on the threats that day. I was very suspicious of going in and out of the airport because it reminded me of Camp Falcon, and I knew a lot of prying eyes usually meant trouble.

It was late afternoon when we arrived, and it was chaotic. At first it wasn't clear where we were going to sleep. But after a few days, and a few moves into different barracks, they finally settled on embedding us with our assigned ISOF companies. I was with the Delta Company again.

All of the terps from Camp Justice moved with us to BIAP, but a few left after only a couple of weeks, spooked by the need to go in and out of the airport. I hated that part of the job, too, and given what had happened at Camp Falcon, I only trusted a handful of terps on my team.

Fahad, the other terp from Baghdad, moved with me from Camp Justice and was assigned to Delta Company, too. I also became close friends with a new terp nicknamed 'Qutt,' even though he was assigned to Charlie Company. Qutt had dark skin and thick, shoulder-length, black hair that reminded me of a rockstar fresh off the stage, effortlessly cool and full of energy. I regarded him as an older brother, and he quickly became my drinking buddy. Qutt had been able to secure a small trailer on base, part of the limited housing offered to some Iraqi officers in Area IV. He lived there with his wife and kids, creating a sense of safety amidst the chaos of war.

Since Area IV was highly restricted, not everybody had clearance to easily go out and come in, so they built little Iraqi food stands for us on base. My favorite one had a small pool stocked with fish. If you wanted grilled fish, they would use a landing net to catch one, gut it, and cook it over a fire pit for you. We often visited these food stands together, sharing meals and strengthening the bonds between soldiers and interpreters, building trust and camaraderie with each meal.

I first met Battalion Commander COL Ali at Camp Justice, but since we hadn't worked together directly, I didn't know much about him. However, once we moved to BIAP, our paths crossed more often because he oversaw all four companies in the battalion. COL Ali was a short man with beady eyes and a thick, dark mustache. He was Shia and had a lot of career ambitions. He led with bravado but lacked real courage. At best, he was a showman. From my first mission with him, I knew I didn't like him.

My company was on a mission cycle when we were briefed on a large-scale operation in Al-Saydia, a predominantly middle-class Sunni area. The mission was to hit multiple targets simultaneously, so several companies would be involved. And Headquarters & Headquarters Company (HHC) would participate, too, because COL Ali wanted to come along. It was considered a low-risk operation, making it safe enough for the commander to join.

But with COL Ali on the mission, it quickly turned into a dog-and-pony show, starting with the mission rehearsal. When it was time to stage, three types of soldiers emerged. Some wore worn-out uniforms and carried dirty, dusty gear—vests, body armor, utility, and magazine pouches. Others, with pressed clean uniforms, were focused on documenting the mission, checking their cameras to ensure they had enough battery life. Then there was the Personnel Other than Grunts (POGs)—who showed up just to be in the photos.

After three to four hours of staging, we finally left the base in a convoy, with the Battalion Commander's Humvee positioned right in the middle.

We arrived at the target area around 2 a.m. And what was supposed to be a straightforward raid quickly became more eventful. As the assault teams began hitting their targets, COL Ali stepped out of his Humvee nearby. Suddenly, the soldiers providing security outside one of the houses came under small arms fire. A man on a rooftop threw a grenade, which exploded near the security team's Humvee.

The soldiers outside returned fire, exchanging a hail of bullets with the enemy. I was inside one of the target houses, listening to the chaos unfolding outside. The radio chatter intensified, filled with urgent orders and updates.

Our mission quickly pivoted to focus on a new target just down the street—we needed to find out who had thrown the grenade. We stormed the house, arrested the men inside, and brought them back to the trucks for questioning and interrogation.

As we began loading up the Humvees to return to base, I walked past COL Ali to get into my vehicle. Just as I was passing by, I overheard him talking to his Executive Officer, or 'OX' for short.

"Dirty Sunnis!" he exclaimed.

At this point, I did not consider myself religious, but I was Sunni, and hearing him generalize a group like that pissed me off. It didn't help that the term 'Dirty Sunni' had been used by the militia and people who were Shia extremists.

I stopped in my tracks and looked at him.

"I am Sunni," I said straight to his face. "Do you want to get rid of me?"

My tone was sharp, and he saw the intensity in my eyes—a mix of frustration and determination that made it clear I wasn't backing down.

"I didn't mean it like that." He tried to backpedal, but it was too late.

Back at the base, we convened for our After-Action Review (AAR), a crucial ritual where we dissected every aspect of the mission. It was a time to celebrate victories, analyze failures, and extract lessons learned from the operation. Afterwards, I left the company HQ and saw a few soldiers standing around laughing about the grenade situation.

I approached them. "What's so funny, guys?"

One of the security guys answered.

"Dude, I saw COL Ali and the XO dive under the Humvee when the grenade hit. They didn't get out until we gave him the all-clear."

I saw the real COL Ali that day. Some commanders lead their soldiers from the front with courage, and some lead from behind with an entourage. COL Ali had shown us his true colors, and I never trusted nor respected him again.

I was still stewing on COL Ali's 'Dirty Sunni' comment when my phone rang, jolting me from my thoughts.

It was my mother, her voice frantic and hysterical.

"Your brother has just been electrocuted!"

"What happened? Is he okay?" I asked, my heart pounding.

She quickly explained that our house had been plugged into the neighborhood generator when the national power grid came back on. Ten-year-old Mustafa had gone to unplug the cable from the generator, but when he gripped the plug, he must have accidentally made contact with the prong, sending a powerful electrical current through his body.

Iraq's infrastructure was destroyed, and we did not have enough electricity. On average, we would have three to four hours a day of electricity from the national power grid. Obviously, living with three to four hours of electricity is very challenging—your food spoils, and your appliances are unreliable. A workaround was to pay for electricity from a privately-owned generator in the neighborhood. In order to get the electricity, you would have to pull a cable from the neighborhood generator and feed it into your home to power the essentials: fridge, freezer, water pump, a few lights, maybe the TV. This would give you an additional eight to twelve hours of power. In the middle of our home, there was a small light that was hooked directly to the national power grid when it was on. That meant we had national power, and when it was off, we had to switch the power to the neighborhood generator. If you looked down the street, you would see a jumble of power cables making almost like a web between different homes.

Dad had heard a strange noise from the generator and went to check it out. He found Mustafa shaking on the floor, still gripping the generator cable. Dad turned off the generator and frantically tried to wake up Mustafa.

"Waleed, we have to get him to the hospital," Mom said, still on the phone. "I don't know if he's going to be okay."

I hung up the phone and immediately called SSG Kris, an 18 Delta medic I often translated for. He was tall and slim, with light brown hair and fair skin. I hoped his medical training might help as I quickly explained the situation, telling him I needed to get my brother to a hospital.

"Do you want to bring him here?" Kris offered. "We can take him to Charlie MED."

I was surprised by his offer. It wasn't standard protocol to have a civilian treated on base, but I was incredibly grateful.

"Of course. Thank you."

Within half an hour, my parents arrived at the gate with Mustafa. Kris and I were waiting there. We grabbed my brother and put him in the back of a Tacoma, speeding towards Charlie MED. I felt a sense of calm, knowing he would receive proper care. When we arrived at the hospital, Kris managed the conversation with the doctor, who held the rank of major. Both the doctor and the nurse were incredibly compassionate and provided excellent care for my brother. After my brother's wounds were dressed, the doctor spoke with me.

"Your brother had pretty severe electrical burns on his right hand where the electricity entered his body and on his stomach where the electricity exited his body. We had to scrub the burnt flesh off his hand and his stomach. He will have pretty bad scars and will most likely need a lot of physical therapy to gain full function in his thumb. But the good news is, he should make a full recovery."

The doctor handed me medications for Mustafa and instructed me to keep a close eye on him, changing his bandages regularly and watching for any signs of infection. With his recovery ahead, my brother stayed with me on base for a week, soaking up the attention. As a ten-year-old boy, he was enamored with the SF guys, their weapons, and, most of all, their gaming consoles.

I expressed my gratitude to Kris for his kind gesture. He had gone above and beyond his duty. He made me feel like a valued member of the team, not just a terp. It was a stark contrast to my experiences at Camp Falcon, where we often felt like disposable assets.

★ ★ ★

This time, my company, Delta, was assigned to serve as the Quick Response team. We were often chosen for the toughest missions due to our high level of competency. This mission came on the heels of a brutal incident. Scouts from the Iraqi Counter Terrorism Force (ICTF) had ventured into Sadr City for reconnaissance. The Mahdi Militia captured two of the scouts and had used their cell phones to call their Sergeant Major, forcing him to listen as they were tortured. We had to react quickly.

Both Delta and Bravo companies were called in to support the ICTF on this raid. It was early evening but not yet dark. Normally, we waited until after dark to utilize our night vision goggles (NVGs), giving us an edge over the insurgents.

My role was to assist the security element in blocking the perimeter road while the ICTF operatives—comparable to US Navy SEALs—executed the raid. The target was a mosque; it was currently doubling as an operations center for the Mahdi Militia and Badr Corps, another terrorist organization.

The ICTF shot the terrorist while he was still in his car, attempting to leave—the same one who had been on the phone with the Sergeant Major. Then, they tossed a thermal grenade, a 2000-degree incineration bomb, into the car. The grenade exploded, fusing his bones right to the car metal.

The raid infuriated the neighbors, who were loyal to the Mahdi militia, sparking a massive firefight. Our company, positioned at the back side of the mosque, was the first to engage.

A civilian emerged from his house with an AK weapon, followed by an older woman, his mother, who was pleading with him not to shoot. Before he could pull the trigger, Firas, who was Iraqi SF, immediately shot him. His mother wailed over his body in the middle of the street.

While ICTF had a lot of action inside the mosque, we were taking a lot of fire outside from civilian sympathizers and insurgents. I was seated in the Humvee behind the driver, positioned diagonally across from Chief Nate, one of new Delta Company advisors, who was monitoring communications. He had brown hair and brown eyes, and his chiseled jawline gave him an air of authority.

Suddenly, I spotted a figure on the second floor of a nearby building wielding an RPK Russian machine gun; he was taking aim at the gunner stationed atop a minigun vehicle that was blocking the opposite side of the road.

Reacting swiftly, I shouted, "RPK, 10 o'clock, second-floor balcony!"

And I opened fire.

Our gunner immediately swung into action, unleashing rounds from the .50-caliber machine gun. Chief Nate grabbed the Squad Automatic Weapon (SAW), braced it on top of the Humvee's hood, and laid down heavy fire. At the same time, the truck-mounted minigun roared to life, unleashing its relentless barrage of bullets in response. The building we were firing at took such heavy damage that the balcony collapsed, sending debris crashing to the ground. Moments later, the power transmitter in front of it exploded into flames, engulfing the area in smoke and sparks.

About fifteen or twenty minutes into the firefight, things started to quiet down. ICTF soldiers began emerging from the building with civilian hostages. There was a female Sunni hostage who had been tortured and raped, and a regular Iraqi Army soldier who had been tortured. There was also an older gentleman who had been kidnapped and held hostage.

Chief Nate looked at me. "ICTF is going to bring a hostage to our position," he said. "Waleed, go get him, and escort him outside the perimeter."

I ran to get the older gentleman as he exited the mosque. He was in utter disbelief.

"May Allah bring you victory over them," he kept repeating. "May Allah protect you."

The ICTF, in a move that seemed a little strange, began gathering the bodies of the fallen insurgents, disposing of them inside the mosque. This action would backfire big time and cause a serious political firestorm for both my brigade and for the US.

I believe the ICTF was also able to retrieve the bodies of the captured scouts while we were attending to the hostages.

A few minutes later, Chief Nate was on the radio, and Intelligence, Surveillance, and Reconnaissance (ISR) just delivered some critical intel.

"How much ammo you guys got left?" he yelled at us.

We all started reporting on how much we still had.

"Conserve your ammo," he directed, his voice cutting through the chaos. "There's a group gathering and heading our way—maybe a hundred or two hundred people. We need to get out of here."

As our Humvee took off, a few militia insurgents started popping up on the roofs of several buildings, firing at us. Our gunners reacted instantly, taking them out one by one as we sped out of the danger zone.

When we got back to base, tensions were sky-high. Both CPT Jassim, who was Delta Company Commander, and CPT Ayad, the Bravo Company Commander, barged into the terp barracks, demanding a meeting with the advisors to protest the ICTF's actions. The advisors quickly redirected them to talk to COL Ali, hoping to avoid any friction. You couldn't blame them for trying to keep the peace.

The aftermath of that mission was devastating. The following day, Iraqi Prime Minister Al-Maliki appeared on national television, denouncing our unit's actions and labeling us the 'The Dirty Brigade.' The local media reported that Iraqi and US Special Forces had attacked people praying in a mosque, which of course outraged the public. But we knew the truth: they hadn't been praying; they were torturing and killing people. But that's not the story the Iraqi civilians heard. It was propaganda at its finest. This mission became known as the Mullah Fiad Mission, and we were branded as the Dirty Brigade.

Some soldiers from Delta and Bravo Company left the battalion, and one of Bravo Company's terps quit as well. I began to see where people's true allegiances lay.

A few days later, I went on leave with Delta Company. Some of us, myself included, headed home. But not everyone made it. There was to be more collateral damage from the previous mission. A group of eight Iraqi SF commandos were kidnapped en route by Iraqi police—who were Mahdi Militia loyalists—and taken to Sadr City, where they were subjected to unspeakable horrors. This was retaliation for being part of the Dirty Brigade. It became clear to everyone that it had been an inside job. The situation had an eerie resemblance to the incident at Camp Falcon. And I realized that I now had much bigger problems.

My terp friend Qutt shared a chilling detail. He had gone on a mission to the morgue to help retrieve the bodies of the eight soldiers. And when they arrived, the JAM was there waiting for the soldiers' family members, so they could capture, torture, and kill them, too.

The incident sparked a battalion-wide investigation, from the highest-ranking officers to the lowest-ranking Iraqi commandos, everyone faced scrutiny. I found myself hooked up to a polygraph, answering questions about everything I knew. I told them about the time I saw CPT Jassim flashing a light in his Humvee during a firefight, signaling to someone not to shoot. And about the gunner that then never fired his weapon. I also told them about the 'Dirty Sunni' comment. There was another concerning piece of information I gave them. I had heard CPT Ayad had been taking home a military-issued laptop containing sensitive information about all his soldiers, as well as intel and maps. He lived in Shu'la, the Mahdi hotspot. I didn't hold back, not when it came to exposing the commanders of Delta and Bravo Company for what they truly were: Mahdi Militia sympathizers.

★★★

The tension outside the wire was extreme, but now I also felt unsafe inside the wire, given the new revelations of my command. I had finished up my leave cycle and needed to get back to base, but I knew, if I made one wrong move, I'd be dead. That day, I rode to the base in a taxi, because

Route Irish was closed due to President Bush arriving at the airport in the afternoon. I decided to just have the cab driver drop me off in a neighborhood close to the airport gate. I figured I could just walk the rest of the way. Big fucking mistake.

I got out of the taxi, and I could see the highway from the street corner where I was standing. There was a mosque on one side, along with a school and a shop. On the other side, there were a few empty lots, with homes under construction.

As I started to walk, I called Fahad. "Hey man, I just got out of the cab. I'm close to the mosque, can you have someone come pick me up?"

"Let me check with Chief Nate," he replied. "I'll see if someone can come get you."

As we were talking, somebody came up from behind.

"Stop. Drop your bag and get off the phone."

"Shit," I thought. "Is today my day?" I brought the phone close to my mouth, and whispered.

"Fahad, I got captured by the mosque. Come get me."

"Don't worry, we're coming," I heard him say, before I slowly hung up the phone.

The man behind me started to pat me down. He took my bag off my shoulder and threw it in front of me. Then he grabbed my wallet and continued to pat me down. He felt my non-military issued Glock 19 pistol strapped to my waist and grabbed it. Luckily, that day I was carrying two pistols. The other pistol was a 9mm Ruger, ISOF issued, in the pocket of my cargo pants. He stepped around me, moving five or six feet in front of me, aiming his pistol directly at my head while tucking mine into the waistband at his back.

My eyes quickly scanned the scene to figure out if I could run. There were two guys with AKs on either side of the street. I glanced up to the rooftop of the mosque and saw a man with a RBK. And then another guy came out of the mosque and blocked the road to make sure nobody was coming. I quickly realized that the men surrounding me must have been hiding in the homes that were under construction on the empty lots.

And as I looked at the situation, I thought, "This is it. I am done."

The man then started rummaging through my wallet.

"Where are you going? Who do you work for?" he screamed.

I quickly came up with a cover. "I'm a guard. I just work here at the gate."

As he was going through my IDs, he saw my Iraqi military ID with a lieutenant rank on it. He continued to interrogate me.

"So you're a lieutenant, huh? I know exactly who you are and who you work for. You're not a guard, you're dirty. You're part of the Dirty Brigade."

"No," I replied as calmly as I could. "I just work here, man. What's going on? What do you want?"

He started to give the men on either side a look, nodding his head as if he was signaling to them to do something. I couldn't play it cool anymore. I had to fight and die or just get captured, get tortured and then die. I figured I might as well fight and take one or two with me.

I slowly moved my hand down toward my cargo pocket while raising my other hand, as I shouted to distract him. My fingers finally slipped into the pocket, gripping the pistol inside. I quickly scanned my surroundings, mapping out my next moves. My brain raced, calculating: first, shoot the guy in front of me, then the two on either side. After that, sprint toward the wall for cover and forget about the guy on the roof. All of this flashed through my mind in about two seconds.

Just then, the man blocking the road started to whistle and signal. The guy in front of me abruptly tossed my wallet and pistol onto the ground.

"Excuse me, dear friend," was all he said, before taking off.

I was baffled. Before I could fully grasp what had just happened, I instinctively grabbed my wallet and gun, cocking it as the men all bolted toward the mosque. Trying to stay unnoticed, I followed them, slowing down as they disappeared inside. I approached the mosque gate with my finger on the trigger, ready. But when I peered inside, it was empty. Deciding not to go any further, I turned around, thinking it was best to get the hell out of there.

I headed left, toward the street that led to the gate and saw a US convoy rolling in. I slid my Glock into my other pocket and crossed the street, making my way toward the gate's checkpoint. That's when I saw Chief Nate driving toward me in a Tacoma. He jumped out of the truck.

"You okay?" Chief Nate asked, his face full of concern.

"Yeah."

What happened? Did you shoot them?"

"No. But I barely made it out alive. I thought today was my day to go until that convoy showed up. The guy at the end of the street must have seen it coming and told the others."

This was my second divine intervention.

"Wow. I'm glad you're okay, dude," Chief Nate said. "You still got your gun?"

I assured Chief Nate the enemy hadn't gotten my issued pistol. And he assured me this incident would be reported.

"We'll talk to the intel guys once we get back. We need to let them know what happened so we can hit that mosque."

Once we got back to base, I gave the intel guys a description of the terrorists and where the mosque was located. A few weeks later another company raided that mosque and found a few of those bad guys.

11

NAVIGATING THE GRAY

Despite the location and the constant fear of getting in and out of the airport, I loved Area IV because we had huge gun ranges for training and a mock city to rehearse our missions. We would train for hours with ISOF, practicing weapon reflexes, marksmanship, mobile target engagement, ambush practice, and fast ropes. My favorite events, however, were the company-wide shooting competitions. We used a lot of ammo, but man, did my aim improve. The range was my happy place. I really enjoyed training with these elite units and learning from them how to be a soldier.

I remember one time a new lieutenant came in who was boasting about how good of a shot he was, so we played a little game using short plastic water bottles as targets. The game consisted of tossing the water bottle down range and then two shooters would start from a low-ready position and shoot at the bottle one round at a time. The goal was to hit the bottle before the other person could. I might have only been a terp, but I smoked him.

★★★

It was all fun and games until it was go time. One day, while we were on a training cycle, Kris burst into the terps' barracks and told me we needed to see the Delta Company Commander right away. I jumped into the Tacoma with him, and we drove straight to HQ. We skipped the formalities and immediately started briefing CPT Jassim.

"Get everyone in the company ready for a mission in thirty minutes," Kris instructed. "There's been an attack on a checkpoint in Al-Mahmoudiya, southwest of BIAP, near a thermal station by the Euphrates River. Two American soldiers were kidnapped, and three were killed. Make sure everybody has enough ammo and water." The temperature was a scorching 115 degrees that day, so water was going to be essential.

We all geared up and headed to the airfield. Four Blackhawk helicopters were en route to pick us up. I remember everyone from the US side was on that mission, along with two Iraqi commando companies—about 120 soldiers in total. The helicopters had to make multiple trips to get everyone to the site.

The Russian thermal station was a failed project halted by the war. You could see some I-beam structures and half-completed buildings scattered around. It was supposed to generate enough power for the surrounding area, including Baghdad, but it had never been completed.

We landed at the Landing Zone (LZ), an open farm right outside the station, and immediately started moving toward the compromised checkpoint. The commandos cleared a few buildings near the gate. Among the structures, we found a white Kia Bongo truck with blood stains on the driver's side door and more in the truck bed. The intense heat had dried the blood, and the Special Forces guys feared the worst. The chances of rescuing the two hostages were dwindling.

We had about half a square mile to sweep. The back of the station ended at the Euphrates River where combat divers began searching the water for the missing soldiers' bodies. During the sweep, we found a weapons cache and a Vehicle-Borne Explosive Device (VBED). Some USSF guys stayed near the VBED while the rest of us made our way back to the station's front gate.

We gathered by the gate, seeking shade, and that's where I met SFC Pete and SFC Dan, both new US team members advising Charlie Company. The moment I met Pete, I liked him. He was an idealist, interested in the big picture and very encouraging. Dan was from North Carolina and very down to earth, but at the same time he'd shoot straight. He was the terps new POC, the man in charge of all the terps, or, as he called it, our babysitter. The three of us bonded quickly, cracking jokes about how unbearably hot it could get in Iraq. The USSF guys had a knack for humor even in the worst situations—a quality I grew to deeply appreciate.

Some USSF soldiers began setting charges to destroy the weapons cache and the VBED. After the detonation, the blast was so powerful that shrapnel injured one of the soldiers. Kris quickly responded, providing immediate care to the wounded soldier and accompanying him back to BIAP.

We were frustrated that we hadn't found any leads on the missing soldiers, but we had done all we could. After the medevac left, we walked back to the LZ, waiting for the Blackhawks to take us back. As we waited, we came under indirect fire from mortar shells that were hitting near the station's front gate—where we had been just minutes earlier.

Small arms fire then erupted where we were, and in an instant, we were caught in a firefight. The helicopters circled above but couldn't land at the LZ due to the intense gunfire below. I found myself next to a USSF guy armed with a .308 rifle, a mid-sized sniper weapon. We were both in prone positions, aiming at anyone with a gun. Suddenly, I saw a insurgent running horizontally from my point of view. I shot at him but missed. Then, I heard a loud bang and saw the insurgent get hit in the side of the head, blood splattering on the wall behind him. The impact was so intense that his head snapped back, as if nailed to the wall, and his feet flew up into the air.

"Dude, nice shot!" I shouted.

After gaining control of the situation, we pursued the shooters for a few hours, going door to door, trying to find out who was responsible for the mortar attack. Each house we raided was a gamble—would we find insurgents or just frightened civilians?

As the hours passed, the lack of water and food took its toll on us. The heat was relentless, and we were running on fumes. In every house we entered, we raided their fridges, drinking any bottled fluids we could find. The desperation for hydration overshadowed any semblance of etiquette. It was a matter of survival.

Night fell, and our resources were nearly depleted. We had little ammo left, no food, and no water. Despite our exhaustive efforts, we found no trace of the missing soldiers.

The houses were eerily empty, with no males to be found, only women and children.

By the time we returned to BIAP, I was beyond exhausted. As I walked off the helicopter, someone handed me a cold Gatorade. I reached for it, but before I could take a sip, I collapsed. I woke up in an ambulance with IV fluids hooked to my arm. Two bags of fluids later, I finally started to feel better.

That day was a harsh reminder of the realities of war. We had done everything we could, but sometimes, even our best efforts weren't enough. The camaraderie and support of my team were the only things that kept me going through those grueling hours.

★ ★ ★

On one particular day we were assigned a big mission. We began by rehearsing our operation in the mock city on base to ensure everyone knew their roles and the location's layout. Our time on target (TOT) was 1:30 a.m., striking when most people were asleep.

We drove our Humvees with no lights, navigating the dark streets to 'Al Shurtah 4th.' As we got closer, different Humvees stopped at their designated spots to lock down the area by blocking the streets. The assault Humvees pulled up in front of the target house. I dismounted, along with the assault personnel, and approached the front gate.

The assault team swiftly breached the door to the house. Once inside, we moved methodically, clearing the house room by room. Precision and coordination were crucial; every corner, closet, and under-bed space had to be checked.

After securing the house, we separated the men and women into different rooms. This was standard procedure to prevent any attempts at resistance or escape. The women, often frightened, were usually more cooperative. The men, on the other hand, ranged from compliant to aggressive, and we had to stay vigilant.

I was in the room with the men, the air thick with tension. The USSF team's 'S2,' the intelligence officer, began interrogating one of the male targets with me as their terp. The questioning was 'nice' by their standards—yelling a few threatening statements and employing some painful interrogation techniques. After a couple of minutes, the suspected terrorist began to talk.

"My wife has the weapons."

I translated and immediately left the room, heading to where the women were being detained.

"Who is that man's wife?" I demanded.

One of the women looked up at me, squatting on the ground and holding an infant in her arms. I walked up to her.

"Your husband said that you have the weapons. I need you to give them to me," I instructed. "And move very slowly when you go get them." The last thing I wanted was for her to dart off to some hiding spot and grab more weapons or detonate an explosive.

She didn't move. She just looked at me, fear etched in her eyes.

"Go get the weapons, slowly," I insisted.

With trembling hands, she reached under her baby and between her legs to pull out a couple of grenades. My eyes widened at the sight. I carefully took them from her.

"Where is the rest?" I demanded.

"That is it," she replied.

I stared into her eyes, hard and unyielding. "If I find more in the house, I promise your husband will disappear. Tell me the truth while you still have a chance."

"I promise," she answered. "That's all I have."

As a terp, I got really good at reading body language. I believed her. The fear on her face was undeniable. I felt a mix of emotions: anger at her husband, for putting her and their infant in such a situation to save his own ass, and heartbreak, for the daily reality women and children faced in this war.

I walked back into the other room and handed one of the grenades to the S2. The other remained clenched in my right hand. I grabbed the suspect by his throat and sucker punched him

"You fucking, coward," I snarled, shoving the grenade into his mouth. "You used your wife and baby to hide your shit!"

The S2 stepped in, removed the grenade, and added it to the evidence. The mission ended with the terrorist being zip-tied, blindfolded, and dragged out to the convoy. We withdrew from the area and headed back to base.

As a terp, I was always caught between worlds, trying to help my people while ensuring the bad guys were identified and caught. Living life in this messy middle had made me begin to question everything. That particular mission marked a big turning point for me. It wasn't just about good guys versus bad guys anymore. It wasn't black and white; the lines blurred, everything was a shade of gray. I found myself in that messy gray area, too, straddling two worlds that were constantly at odds, where right and wrong were no longer clear-cut but instead tangled up in fear, survival, and loyalty.

Sa'ad had been gray. The woman clutching her baby while holding a grenade was gray. They were both trapped, forced to do things they didn't want to do. That's the brutal reality of war. Innocent people caught in the middle, coerced or forced into making terrible choices.

I'd seen kids as young as fourteen shooting at us during a raid, men using women and children as human shields or as suicide bombers. Who in their right mind would make their wife and baby sit on a grenade? I was angry at the cowardice of the terrorists. I wished they would just pick up a weapon and fight back man to man instead of using the innocent and vulnerable.

The complexity and moral ambiguity of war weighed heavily on me. I was no longer a naive college kid; I was living in a world where every decision carried immense weight. Balancing empathy for those trapped in impossible situations with the demands of my role became an emotional tightrope. Navigating the murky waters of right and wrong, I struggled to protect the innocent while dismantling terrorism. Every day tested my resolve, forcing me to see beyond the simplistic black-and-white narratives that we so often crave. War wasn't just happening on a battlefield around me—it was a constant battle within myself, a relentless struggle to reconcile who I wanted to be with the person I was becoming.

★ ★ ★

It was finally my leave week again, and I was ready for a break from the chaos. I was still trying to wrap my head around that baby and the grenade. And when I arrived home, Rana wasn't there. I called her to see where she was.

She was at her parents' place, of course. I made my way over to their house, and when I walked in, I could see Rana lying on the bed. Quickly, I moved towards her and knelt by her bedside.

"Rana, are you okay?" I looked around the room, confused. I didn't know if she had just had a fight with her mother or if she was sick. But I could sense something was wrong. The tension was palpable; it felt like the room itself was holding its breath, waiting for something to break.

"I'm just not feeling well," she finally said.

"Have you gone to the doctor?" I replied, as I felt her head to see if she had a fever.

"No. I think I just need some rest."

I let her rest for a few hours and just waited until she started complaining of severe stomach pain. That's when I got really worried.

"Rana, let me take you to the clinic," I began to insist. "Just to make sure it's nothing serious."

She didn't respond. I looked over at her mother, expecting her to back me up and encourage Rana to see a doctor. But for the first time, her mother was oddly silent. She didn't seem concerned at all. Rana let out a whimper of pain, and that was it—I made up my mind.

"We're going to the doctor," I said firmly.

After some convincing, I finally got Rana to agree and helped her to the car. I drove straight to a nearby clinic, and we waited in line for what felt like an eternity before a doctor finally came to see us.

"When was the last time you had your period?" the doctor asked.

"I'm not sure, I don't really remember."

The doctor checked her out.

"It looks like you just had a miscarriage," he said. "Were you aware you were pregnant?" Rana shook her head no.

I was in shock. I had no idea she was pregnant, and I felt a deep sense of loss, knowing that our baby died. I just kept picturing that little baby with the grenade. I had always wanted to be a father. And yet, you have fathers out there that fucking make their babies sit on grenades. I couldn't shake this thought. I was devastated and confused.

"Go home and get some rest," the doctor said kindly. "You will be okay, and you can try again soon."

I drove Rana home in silence, my mind numb and my heart aching. The weight of what had happened hung heavy between us, like a wall neither of us dared to breach. I kept glancing over at her, hoping for a sign that we could talk about it, that we could somehow make sense of the pain. But her eyes were distant, staring out the window, lost in a world of her own.

When we pulled into the driveway, I turned off the car and sat there for a moment, struggling to find the words. Nothing came. She quietly got out of the car and walked inside without looking back. We never spoke about our baby again.

12

CIVIL WAR — SUNNI VS. SHIA

I decided it was time to throw my frustrations into the aggressive work of an assault team, and almost just as quickly, my country faced assault from another enemy. Suicide bombers attacked the Al-Askari Shrine in Samarra, some seventy-five miles from Baghdad. Everyone was in shock.

The attack caused confusion and deepened sectarian hatred. Some believed Iraqi Sunni Muslims were responsible for targeting this holy Shia site, but that didn't add up. Iraqis typically didn't use suicide bombings nor did they discriminate between Islamic sects. Many suspected Iranian militia involvement that aimed to plunge the country into further turmoil—a tactic that would eventually succeed in sparking civil unrest between Sunnis and Shias.

The nation was a powder keg, and Baghdad was at its heart. It had been a melting pot of all the tribes. And now, as the violence escalated, the lines between friend and foe blurred. Our once-united neighborhoods became divided battlegrounds, each day bringing new threats and uncertainties. And I watched as my country began to unravel into civil war.

Late one afternoon, my phone rang. It was Rana. She was frantic. Screaming.

"There is a lot of activity in the neighborhood," she wailed. "I'm not sure what is happening, but I can hear a lot of gunshots."

"Don't worry," I told her. "I'll come to you."

"But I think the neighborhood is on lockdown. No one can come in or out."

"I'll figure it out. I promise," I said, as I hung up.

There had been relentless fighting between Ghazaliya, a historically Sunni area, and the adjacent Shia neighborhood. The sectarian violence was at its peak, with the US estimating that an average of 15,000 Iraqi civilians were dying violent deaths each month. Almost half of these fatalities occurred in Baghdad. My city was drenched in the blood of innocent men, women, and children caught in the crossfire. Extremist groups like Al-Qaeda in Iraq clashed with Iran-backed militias like the JAM, who sought ethnic cleansing. The conflict had shifted from America versus Saddam and his regime to neighbor against neighbor. It was becoming a genocide of epic proportions, but it wasn't a simple matter of good versus evil or truth versus lies. The situation was so complex that, most of the time, you couldn't even be sure who the real enemy was.

Given this uptick in violence, I was terrified that Rana was in danger. I left the base carrying my pistol and jumped into a taxi. As we approached Ghazaliya, the sound of gunshots filled the air. The main entrance to our neighborhood was indeed blocked off, so I asked the driver to drop me off a few streets away, towards the back of my neighborhood. I quickly got out of the cab and looked around to get my bearings. Siddiq Street, which cut through the neighborhood from east to west, was empty. I crossed the street, heading east towards my house. As I passed the Al Siddiq mosque, a couple of rounds hit the ground right in front of me.

Shit. I knew those were warning shots. I turned and walked down a street opposite the mosque. When I heard Humvees approaching, I started to run. It had suddenly occurred to me that I had to get off the street or I was going to get killed.

The gunners on the Humvees began engaging the insurgents, the loud cracking of the .50-caliber machine gun piercing the air. The firefight intensified, bullets flying in all directions. To my right was the mosque with its tall walls; to my left, a house with a front wall covered from top to bottom in bushes. I dove headfirst into the bushes, seeking cover.

I was dressed in all black and carrying a pistol, making me look suspicious as hell. I just hoped the Americans wouldn't spot me because I looked exactly like the enemy. The heavy gunfire continued, each second stretching into what felt like an eternity, but it probably lasted only ten minutes. When I heard the Humvees start moving again, I waited a few more minutes, then climbed out of the bushes.

I took off running toward my house, moving as fast as I could. As I sprinted, I saw two cars riddled with bullet holes. One of them had bodies slumped over, hanging out of the windows. Blood pooled around the car, shoes floating in the red trails. Panic surged through me, and I pushed myself to run even faster.

As I turned down my street, I heard the unmistakable sound of an Apache helicopter hovering over the neighborhood. It had probably been providing air support during the battle, but now its attention was locked on me. I was the only person running down the street, making me an obvious target. My pistol was strapped to my waist, so I looked up and raised one hand as a universal signal not to shoot, trying to show them I was friendly. I slowed my pace to a walk, keeping my hand in the air, and continued like that until I reached my house.

My adrenaline was still surging as I pushed open the door. Rana rushed toward me, throwing her arms around me.

"I was so scared, Waleed," she said, her voice shaky. "It feels like every day could be our last."

She was right. The fighting had become worse and was more violent. And I tried to reassure her that we would make it through. But despite my presence, the tension between us remained.

The US tactic was to roll through neighborhoods and draw fire. This was successful in one way, because the insurgency would come out and fight, allowing the US troops to capture or kill them. But on the other hand, anyone who was out on the street was considered the enemy and was at risk of being shot.

A few days later, my phone rang. It was a US number. Luckily, I was home and not out and about, or I wouldn't have been able to pick up.

"Hello," I answered.

"Is this Waleed?" asked the man on the other end. The voice sounded familiar.

I hesitated but then said, "Yes."

I didn't expect anyone from the States to call me—all my US friends had Iraqi cell phone numbers.

"Hey buddy, it's Kris, the medic, calling you from America."

"What's up, brother? How's it going?" We talked a bit.

Kris had been home for a while by then, and he asked me about the Delta Company guys and the terps.

Then he said, "Listen, I created an email for you and attached a few documents to it." He spelled out the address for me and gave me the password, which was 'toshiba.'

"Once you're logged in, you can change your password, and I'll communicate with you using that email."

I didn't know what to expect. At that time, I didn't have internet at home, but there was an internet café nearby with a bunch of computers you could rent. While Rana was resting, I left the house and walked to the shop.

The room was cramped, lined with about four computers on either side, all facing inward. I chose one near the far end, hoping for a little privacy. The hum of the machines filled the silence as I typed in the familiar web address: *www.yahoo.com*. Time seemed to drag as I waited for my new email account to load, my heart pounding in anticipation. At last, the inbox appeared. There it was—a draft email with a series of attached documents.

I opened the first one and saw it was an immigration form. I immediately closed the document and deleted the browser history. I quickly paid for my session, anxious to leave the shop, worried someone might have seen the document. Reading a document in English with a US government crest on it could literally get me killed in my neighborhood if the wrong person saw it.

By the time I got home, I was full of questions. What was in that document? Since I knew I had internet access on base, I decided it would be safer for me to wait and check it there.

A few days later, back on base, I opened my computer and started going through the forms. There was a Word document Kris had written, explaining that the US government had started a Special Immigration Visa (SIV) program for US allies, interpreters, and their families to immigrate to the United States. He provided some instructions, offered to sponsor my visa, and even offered me a place to stay once I arrived.

I was in shock. What? I leaned back in my chair and exhaled deeply. I had never known there was a way out. In my mind, I had just accepted that I would live and die by the gun, thinking I'd never grow old. Suddenly, it hit me: there actually was a way out. This was my golden ticket. Freedom felt so intensely valuable, considering all I had sacrificed.

My heart pounded as I dialed Kris's number, eager to hear more details about the visa. He picked up. His voice was a lifeline, and I couldn't thank him enough for remembering me and going through all the effort to help me. This was the first step. The next would be convincing Rana to come with me.

★ ★ ★

A new Captain with US Special Forces Fifth Group had rotated in, CPT Nelson. He had strawberry blonde hair, a thick mustache, and a prominent nose. On this particular day, CPT Nelson's usual translator was unavailable, so he came into the barracks looking for another terp.

The door to my room was open, so CPT Nelson looked in. "Hey, can you come help me out here?" he asked. "Looks like there is an issue at the gate."

I followed him to the base gate, where a convoy full of Kurdish soldiers from the battalion was waiting to head north for their leave week. The American guards at the gate were raising concerns, saying the convoy was carrying far more ammunition than it should—many more rounds than were allowed.

Bravo Company Commander CPT Ayad, who was in charge of the convoy, gave me a serious look. From that one glance, I understood he expected me to mistranslate, making sure the conversation went in favor of letting the convoy through. From the look I gave him, he knew immediately that I wouldn't go along with his nefarious plan. He did not like that at all. Just then, COL Ali arrived. And I knew he would have no problem putting CPT Ayad's spin on the situation.

I turned to the convoy and said what CPT Nelson told me to translate.

"You guys aren't supposed to take that much ammo with you. This is way more than we even take out on a mission."

CPT Ayad's response back to me was so disrespectful that it almost led to a physical fight. COL Ali had to step in and break it up. That was the moment when tensions really started to escalate between CPT Ayad, COL Ali, CPT Jassim and me. It became clear that COL Ali's corruption, racism, and favoritism were a pattern. And those in his inner circle, which included CPT Ayad and CPT Jassim, were no different. They were like vultures, ready to exploit every opportunity for their own gain.

CPT Nelson ordered COL Ali and the convoy commander to return some of the ammo to the arms room. CPT Ayad was furious; his plans had been thwarted, and he couldn't carry out whatever scheme he had in mind.

That's how we found out that some of the battalion soldiers were smuggling weapons and ammo to support the Peshmerga militia up north. Peshmerga, meaning 'Those Who Face Death,' was organized in the 1940s as the standing army of Kurdistan, but soon it became a guerrilla force outside government control. Various Peshmerga units were loyal to various factions in Kurdistan, but none of them were loyal to Iraq, even though Iraq had control of half of Kurdistan since World War I.

Nothing happened to CPT Ayad as a result of this incident. But it upped the stakes for me. I was already on their radar as noncooperative after the polygraph test, and now I was sure they were watching my every move.

By now, CPT Nelson had seen enough corrupt activity that he was ready to investigate, so he asked me about it.

"How confident are you in what you've heard?" he wanted to know. "Do you know about other stuff for certain or is it all just rumors?"

"I don't have anything documented. But I'm hearing plenty from the soldiers," I told him.

"Let's continue this conversation," he said. "I have someone you need to talk with."

He arranged a meeting for me with an investigator. We met at a coffee shop on the larger base outside of Area IV. As we sat down, I shared that I'd been hearing a lot of talk from soldiers about ammo being smuggled up north and other troubling activities happening within the company.

I eventually found out that the investigator 'Johnny' was part of another government agency. He asked me if I could find anyone else willing to testify in person or by written statement.

"I'll see what I can do," I replied.

Over the next week or two, I gathered the information he needed. I spoke with some of the soldiers who had mentioned those allegations.

"If you're fed up and want to talk, I can connect you with people who will handle this," I assured them.

I promised them the same thing Johnny had promised me—that they could remain anonymous, that their identities would be protected. In the end, about eleven or twelve soldiers agreed to write statements.

We organized a covert meeting at 2 a.m. in the interpreter barracks. To maintain secrecy, we sent three soldiers off base every fifteen minutes in a blacked-out van to write their official statements against COL Ali, CPT Ayad, and CPT Jassim. Each trip was carefully timed to avoid suspicion, each soldier ready to reveal the corruption they had witnessed.

The Delta soldiers were deeply divided. Some were pissed off about the eight commandos who were killed on leave after the Dirty Brigade incident and blamed COL Ali, while others were fiercely loyal to him, even spying for him. I started meeting pretty regularly with Johnny, telling him about the concerning things happening on base and all the information I disclosed during the polygraph test.

No big surprise: COL Ali heard about it anyway. He had plenty of ears keeping him informed. But it didn't take me long to figure out who was feeding him information. I used a technique I'd picked up from Johnny. I'd shared different bits of false information with the people I suspected were his informants. Sure enough, COL Ali heard some of those lies, and I could identify the informants based on which lies reached him.

When he confronted me, I said, "I just wanted to find out who was sharing these lies, and now I know because I know who I told them to."

A flicker of surprise crossed COL Ali's face, quickly replaced by anger.

"You're playing a dangerous game," he warned, leaning forward. "This isn't a joke. You're supposed to be on our side."

"I am on the side of telling the truth," I shot back.

From that point on, I was even more at odds with COL Ali and the other Company Commanders. The translators were also divided, and some of them felt that I had put them in a difficult position. Several of them stopped talking to me and distanced themselves completely. The dynamics in the barracks became really intense. But I was hard-headed and refused to let my concerns go. They, however, just wanted me to be quiet. And the HHC translator did not like that I was talking about concerns instead of just translating.

"Keep your mouth shut," he told me. "This is not what you're here for."

I knew one of the Bravo Company translators, Mike, wanted us to reconcile and move on.

"COL Ali wants to talk with us, all the translators together," he announced one day, calling for a meeting in COL Ali's office.

At this point, I didn't have much hope that anything would change, but my curiosity got the better of me. Mike had organized a big meeting with all the translators, and COL Ali was to play the mediator. But instead of finding common ground, the tension in the room escalated. COL Ali couldn't resist making a snarky remark in my direction. Leaning on one elbow instead of sitting up straight like an officer, he smirked.

"Causing a lot of trouble, aren't you?" he said.

I snapped back. "I wouldn't cause trouble if there wasn't any. There are things that are wrong, and they need to stop. That's why I speak up."

That exchange made it clear that there was no repairing my relationship with the translators who sided with COL Ali, and honestly, I didn't care anymore. I was focused on seeking justice, no matter the cost.

One afternoon, CPT Jassim confronted me outside the barracks. His face was flushed with anger as he cornered me.

"You're supposed to be on our side!" he barked.

I held my ground.

"I am on your side. I'm just not on the Mahdi militia side," I said firmly. "You do your job, and I'll do mine."

This moment marked a turning point where the true allegiances within our battalion became unmistakable. The silence that followed was thick with unspoken threats. From that day on, I constantly felt the weight of disapproval from my colleagues; I felt that their eyes were always on me. But I refused to back down. For me, this wasn't just about the job; it was about doing what was right and protecting the integrity of our mission.

"Justice isn't always popular," I'd remind myself. "But it's necessary."

In what would become my last meeting with Johnny, he took me to the Radwaniyah Palace, the sprawling complex where all the top US brass operated. The palace stood at the southwestern edge of BIAP, offering a commanding view of the airport control towers, the runways, and the surrounding properties that once belonged to Saddam. The opulence of the palace, with its marble floors and towering columns, contrasted sharply with the grim reality of the war raging outside.

Johnny pulled me aside, he tried to be reassuring.

"Don't worry about COL Ali," he said quietly. "We'll take care of it."

But moments before, while I was approaching, I had heard an Army commander speaking to Johnny. His voice was low, but it had carried an edge of frustration.

"Getting rid of COL Ali would be pointless," the commander had said. "The whole culture is corrupt. We take out COL Ali, and we'd have to take them all out. It's just their culture."

Indignant anger had flared inside me. And after Johnny's attempt at reassurance, I couldn't help but let it out.

"Just for the record, corruption is not our culture."

Johnny looked at me, his face a mix of concern and helplessness, but he said nothing. We walked back to the car. The drive back to the base was tense and silent. I stared out the window, seething. Yes, there were corrupt people in Iraq, but that didn't define us all. I thought of my family, my friends, and the countless Iraqis risking everything for a better future. We weren't all corrupt.

Outside the wire, my country was tearing itself apart in a civil war, and now I was fighting my own battle on base that seemed hopeless. The sense of betrayal gnawed at me. How could we rebuild Iraq when the very people meant to help us saw us as inherently corrupt?

13

THE SURGE

In January 2007, President George W. Bush announced a bold new strategy called 'The Surge' in response to the extreme sectarian violence and rapidly deteriorating conditions in Iraq. The surge called for an increased military presence on the ground; around 30,000 additional US troops, including five additional brigades, were being deployed. Tours were being extended for many soldiers. The idea was to surge into the community, to protect locals and eradicate extremists.

The change was immediate and dramatic. Troops were no longer confined to large, fortified bases. Instead, some moved into neighborhoods all across Baghdad, turning bombed-out houses into Combat Outposts (COPs). The goal was clear: embed soldiers into the communities to quell the violence from within.

The first six months of the surge were the deadliest on record since the invasion. Both Iraqi civilians and US soldiers faced unprecedented levels of violence. The streets of Baghdad were becoming bloody battlegrounds, riddled with insurgent gunfire, suicide bombers, VBEDs, and now a new weapon developed by Iran, called Explosively Formed Penetrators (EFPs) that could shred a Humvee like a hot knife through butter.

Around this time, CPT Jassim was finally removed from his position as Delta Company Commander and replaced by MAJ Hassan. An experienced Iraqi Special Operations Forces veteran, MAJ Hassan had been through military school and seen a lot of combat. He was an outstanding military leader—decorated, respected—and his units had often earned 'Best Company' honors. Even other officers admired him.

Unlike his predecessor, MAJ Hassan was a man of integrity who refused to bow to COL Ali's corrupt influence. However, his defiance came at a cost. Under MAJ Hassan's leadership, Delta Company soldiers faced unfair treatment as COL Ali sought to undermine them out of spite. Despite this, MAJ Hassan remained steadfast, determined to uphold his principles, even in the face of relentless pressure from the top.

Also around this time, the US sent in a new team. Usually, when a new team arrives, the prior team makes some pretty formal introductions, and the whole meeting is pretty stiff. This team was different right from the start.

The outgoing advisor said to me, "This is the guy who will be taking my position. His name is SFC Wil. Wil, this is Waleed."

Wil jumped right in. "We've got the same name. I'm going to call you Willy from now on."

We connected instantly.

"Cool," I said. "I like it."

Then we went to meet with MAJ Hassan. From the start, he, Wil, and I formed a great bond. We just clicked. We began doing extensive training sessions together, and Wil taught us a lot. Wil was also incredibly fun to be around.

Unlike some advisors who carried a lot of frustration from their previous deployments, Wil brought a fresh and positive energy. He set up competitions among us during training, which the guys really enjoyed. His presence not only improved our skills but also boosted our morale, making Delta Company a cohesive and formidable company.

One of my first missions with Wil turned out to be quite amusing. We were raiding some houses to find insurgents. Traditional Iraqi houses were set back from the street, surrounded by a wall. To enter, you had to go through a front gate, which was the only opening in the wall.

Once inside, you generally faced two front doors right next to each other. The larger wooden door led into the main part of the house. Next to it was a smaller door, which opened directly into the living room or guest room—the nicest room in the house. There was also a side entrance that took you to the kitchen at the rear of the house.

As we approached the house, I yelled to the commandos, "I'm with Wil!"

Wil normally carried numerous breaching charges so he was always ready to breach a structure. As soon as we entered the courtyard leading to the front doors, Wil moved toward one of the doors and began placing his charge. He had the charge set and ready to blow just as I noticed the Iraqi SF commandos entering the house through the kitchen door.

I followed the commandos into the house, expecting Wil to be right behind me. I assumed Wil knew the commandos were already inside. I circled around to the open kitchen door, planning to go through the house and open the front door from the inside, so we could clear the building from back to front.

I made it to the living room and was walking toward the foyer to the front door when, suddenly, someone grabbed me by the Kevlar plate on the back of my body armor and yanked me away from the entry. A split second later, I was hit by a wave of intense heat and a burst of pressure, like a powerful kick to the chest.

Kaboom! The front door blew open.

I turned around; it was Wil behind me.

"Man, you almost blew me up!" I shouted.

"I told you I was going to blow it!" he yelled back.

"The door was already open."

"I didn't see the door open. I told you guys I was going to blow it." Wil insisted. "I don't want to carry these charges around with me all night. Each house gets a charge, that's how we scare the bad guys."

All this happened within seconds, but we kept our banter going back and forth for a bit, laughing at each other because it had been such a close call. In the middle of these tense situations, sometimes we just needed to laugh. Like a release valve for all the built-up pressure, a moment of levity helped in the midst of chaos.

Missions with Wil were never dull. There was a sense of excitement in the air whenever we teamed up, a feeling that anything could happen—and it usually did. But amidst the chaos and uncertainty, one thing was certain: Wil would always have my back. And my front if I wasn't careful.

★ ★ ★

I paced back and forth in the living room, my heart pounding with a mix of nervousness and excitement. I had to tell Rana about the visa. Deep down, I knew convincing her wouldn't be easy. But still, I had hope that perhaps she would see this for what it was: a fresh start. A ticket to freedom for both of us.

I decided it was time to say something.

"Waleed, what is it?" she asked, as I continued pacing. "You are making me nervous. Can you just sit down?"

I exhaled and slumped onto our worn-out sofa, searching for the right words. After a long pause, I finally let them out.

"I have something important to tell you," I started, my voice trembling slightly. "Something that could change everything for us."

Rana's smile faded, replaced by a cautious expression. "What is it?"

"I've been filling out paperwork for a Special Immigration Visa—for us to go to the United States. If everything goes through, we could start a new life there. What would you think about going to America? Away from all of this, just you and me?" I asked, my voice filled with excitement, hoping she could see the potential in this opportunity.

Rana's eyes widened, and for a moment, there was silence. I watched her closely, waiting for the words to sink in. Rana's expression turned to shock and then anger.

"What?" she said sharply. "You want me to leave my family?"

My heart sank.

"I thought you'd be happy. This is our chance, Rana. Things aren't getting better here—they're only getting worse. We're still in the middle of a civil war. And I want a safe and free future for us and for our future children."

Rana shook her head, her eyes filling with tears.

"I don't want to leave," she said, her voice breaking. "This is my home, Waleed. My parents are here, my culture, everything I know. How can you expect me to just abandon all of that?"

I put my arms around her, assuring her things would be okay.

"I don't want to leave my family either," I replied gently. "But Rana, it's the only way. I'm trying to save us, to give us a chance at a life where we don't have to live in fear every day. Don't you see that?"

"No, Waleed," Rana shot back, anger flaring in her voice. "I can't just leave my parents, my home, my country because things are hard!"

The excitement I felt only moments ago was now replaced by a heavy sense of dread.

"I thought you'd understand," I said quietly. "I thought you'd want this, too."

"Well, I don't," Rana replied, tears streaming down her face. "I don't want to go. I'm not leaving my family!"

"Rana, you are my family now!" I tried to reason with her, but my words fell flat. Her eyes were distant, and instead of reassurance, I felt the tension between us growing, the divide widening.

I could see I wasn't getting through to her. So, I decided to let it go for now, hoping she'd come around in time and realize the bigger picture—that this visa was our only hope for a future, our chance at survival.

★ ★ ★

One day, Wil came into my room with a mischievous glint in his eye.

"Hey, Willy," he said, "I'm heading out with Charlie Company on a special mission. Want to join?"

"Absolutely," I replied without hesitation.

"It's gonna be different from other missions, but I think you'll enjoy it," he teased.

"I'm all ears," I said. "I'm ready for whatever."

Together, we left my room and made our way to Charlie Company for the mission brief.

The mission was to capture a US national of Iranian descent. According to a confidential informant, this individual had been involved in some shady dealings, from laundering counterfeit US dollar bills to alleged kidnappings and killings. The mission brief was intense, with Iraqi soldiers bombarding the advisors with questions, especially about the Rules Of Engagement (ROE). Shoot if shot at—simple enough, but the Iraqis needed reassurance, as this is their first time going after an American target.

As we geared up to head into the Red Zone, our target area, our first stop was at the iconic Crossed Swords Park. The statue of the crossed swords, built by Saddam Hussein after the Iraq-Iran war, symbolized victory. We used this landmark as our staging area, where other government agencies would join us for the mission. While we waited, we took a few photos, soaking in the moment. It was surreal—aware of the dangers ahead, yet appreciating the camaraderie and the memories we were making.

Once it was go time, we wasted no time in taking control of the target house. But alas, our American target was nowhere to be found. Undeterred, we expanded our search to neighboring houses, following every lead. That night, we ended up clearing at least nineteen houses. We ran out of all of our explosive charges and ballistic breaching rounds, so we had to start breaching doors using a sledgehammer and a Halligan tool—the tool firefighters use to break into burning houses. We were exhausted.

During a pause in the action, Wil strolled towards us carrying a massive bronze sword. Our ensuing exchange was classic.

"Hey, Willy," Wil said, brandishing a massive ornate metal sword.

"Holy shit, dude! Where did you find that?"

"I found it somewhere. I need you to take this for me."

"How am I going to do that?" I protested.

I don't know, shove it in your pants or something."

"What? I can't do that. I'll be waddling like a fucking penguin!" I said, laughing.

"Nah, you'll be good, you'll see."

So Wil and I started to shove this massive sword into my pants, as I continued to argue with him that this would never work.

"Stop being a bitch," he laughed. "Just put it in your pants, it's going to work."

"Dude, even if I tried, that thing would stick out all the way to my shoulder."

Once the sword was in my pants, it literally reached up to my shoulder. I was correct.

Wil looked at me, "See, it's gonna work."

I laughed. I didn't know if he was trying to convince me or himself.

And so it went. Time and time again our bantering turned stressful situations into moments of camaraderie and even laughter. While some succumbed to fear, others seized the opportunity to turn adversity into adventure. Wil exemplified this spirit.

Needless to say, we returned to base without the sword but we had stories to tell. I couldn't help but be grateful for brothers like Wil. In the Special Forces community, trust was everything, and he was one of the few I'd follow anywhere. Wil's assignment was extended and overlapped with a couple of US teams, so we had the pleasure of working together for an entire year.

One day, we received a mission to hit a target in Al-Shu'ala, one of the most dangerous parts of Baghdad, second only to Sadr City. The mission was focused on capturing a member of the Mahdi militia. During our mission brief and rehearsal, we were told it was a 'Time Sensitive Target.' This meant that an informant in the target area was keeping an eye out for our person of interest. As soon as the target appeared, we'd get a call and would need to head out immediately to capture him.

Our Humvees were lined up and ready to go, staged just outside the company HQ. I was going over the mission details with MAJ Hassan and Wil. We were sipping hot Iraqi chai, adrenaline running high because we all knew this was Wil's last mission before he had to fly home.

The call came: The target was in his house. We all rushed to the Humvees, and somehow the third squad leader Bassim got into the spot where I was supposed to sit.

"That's my seat," I protested.

"No, it's mine," Bassim insisted.

He and I were still arguing about the seat when the first Humvee started moving out. So I ran to the next Humvee in the convoy and got into the seat behind the driver.

As we went into Al-Shu'ala, the new SF lead navigator—new to the job and also new to the area—made a huge mistake. We ended up driving around for at least thirty minutes. On a typical mission, we're in and out in fifteen minutes, so this drew a lot of unnecessary attention. It was maybe one thirty or two o'clock in the morning, and as we drove from one area into another, we knew everyone could hear the Humvees in the dead silent streets. Random shots began ricocheting off the Humvees, but we couldn't identify where the shooters were hiding.

We finally made it to the target building. Wil blew a charge on the front door, and we rushed inside. There was hot chai on the table, but the house was empty. The target must have left just moments before we arrived.

We had been warned that a bunch of militia were coming our way, and we needed to get off the streets quickly. Moving single file through the neighborhood, our convoy finally made it all the way back out to the main street. I was still in the third Humvee, looking through the side window and the front windshield, keeping an eye out.

Kaboom! A massive explosion lit up the night sky as it hit the Humvee right in front of us. Our driver instantly veered left, swerving around the damaged Humvee, and then pulled up ahead to provide cover. An EFP had gone off, and standard procedures kicked in automatically; all our training took over. Gunshots erupted all around us, filling the air with chaos. Some of the shots might have been from our own guys, but no one could tell where they were coming from.

All the guys in our vehicle jumped out and rushed toward the Humvee that had been hit. It had rolled over to the right side of the road where it had come to a stop. I reached the vehicle and peered inside.

The JAM had used an EFP. We were familiar with IEDs and had learned to avoid potholes because insurgents would often bury IEDs in them, covering them with dirt.

IEDs exploded outward and upward, but EFPs were different; they were directional and much more deadly. These bombs could be triggered remotely with a cell phone or by a wired trigger. This EFP had been placed on the side of the road, and we knew whoever had set it off had to be nearby. Later, we would discover that three EFPs had been daisy-chained together, all aimed at the road on which we had been traveling.

The EFP had sliced into the Humvee from the right side, cutting Delta Company's Sergeant Major Badr, who was sitting in the right front passenger seat handling communications, in half. Unfortunately, he chose to join this mission—Wil's last—rather than taking leave to prepare for his wedding. The driver had been hit in the neck, the gunner had lost his leg, and the soldier behind Badr had been hit in his right arm. I opened the back door behind the driver—the seat I was supposed to be in—and saw Bassim, his head leaning backward, motionless.

"Hey, Bassim, you good?" I asked.

No response.

I swung my rifle behind me, letting it hang from its sling, and reached into the Humvee with my left hand to grab Bassim.

My hand went straight through his body.

He'd been hit in the right side of his torso, and there was a large, gaping hole where the EFP had torn through his body. His eyes were rolled back, and I could tell he was already gone. I grabbed Bassim by a part of his body armor that was still intact and pulled him out of the Humvee.

That was it for me. I didn't even check on the driver. I just grabbed my rifle and started firing. People were running down the street, and at that hour, I knew nobody was out for a casual stroll. I just aimed and shot, aimed and shot, aimed and shot.

Rounds were going off everywhere, bullets flying in every direction. Wil was beside me. After a few minutes of intense engagement, he turned to me.

"Tell the Iraqi commandos to stop using their rifles until the .50-caliber machine gun runs out of ammo. Once that happens, our rifles will be all we have left for protection."

We called for the Quick Reaction Force on comms, but the area was so hot that no one was coming to extract us. Later I learned that Qutt had heard over the radio that Delta had taken a hit, resulting in two Killed in Action (KIAs) and three wounded. His heart sank, fearing I was one of the casualties, and he started worrying about how he'd break the news to my mother.

Luckily, Charlie Company was nearby, heading to another mission. They immediately canceled their mission and turned around to come to our aid. As soon as they arrived and we'd gained control of the fighting, we grabbed litters and started loading up the bodies.

Then we tried to exfiltrate and recover the disabled Humvee. We managed to drag it to the end of the street, but it was beyond repair. We waited far longer than we should have for a tow, leaving us exposed. We were supposed to have been out of the hot zone three hours earlier. By now it was dawn, and we were getting even more vulnerable in the daylight. Morning traffic began to stir around us, increasing the tension as civilians started appearing on the streets.

Those of us tasked with security while waiting for the tow were seething. After what happened, we felt justified in doing whatever we deemed necessary to protect ourselves.

Just then, a big, old, blue Mercedes dump truck started approaching us. We were on high alert, stopping anyone from getting too close, uncertain of what threats they might pose. An Iraqi commando squad leader stepped forward, waving his arms, signaling to the truck to turn around. But it kept coming.

"Stop!" I yelled. "Don't shoot that guy!" my voice trying to cut through the chaos, desperate to prevent another tragedy.

But the tension was too high, frustration boiling over. Someone fired a shot.

"Stop!" I screamed. "He could be a father."

Just then a bullet hit the driver, and the truck screeched to a halt, its horn blaring as it came to a stop.

As we headed back to base, adrenaline still surged through my veins. My mind was spinning with thoughts of Bassim. He had taken my seat in the Humvee. If I had been in my usual spot, I'd be the one dead. The weight of that realization hit me hard, filling me with guilt. Why Bassim and not me? Survivor's guilt clawed at my conscience, the burden of knowing that someone else had paid the price for my safety. Yet, alongside the guilt was a profound sense of gratitude. I was still alive and breathing; my heart was still beating. It felt like another close call, another divine intervention that had spared my life.

The next day, we gathered at Qutt's trailer. He had sent his wife and kids to visit their grandparents, giving us space to decompress. Two refrigerators were stocked with beer, and there was no shortage of whiskey and vodka. I escaped into music, just like I had when the war first began, blasting Linkin Park as we drank one bottle after another, drowning our grief and rage.

The guilt I felt about Bassim's death began to shift, hardening into anger. Why him and not me? The unfairness of it all seared in my mind, and my sorrow twisted into a desire for revenge. We were all pretty wasted, but there was a unifying, burning focus: avenging our friends. Our conversation circled around the same thought over and over —the need to hunt down and destroy the militia responsible. It wasn't just about grief anymore; it was about justice, retribution, and making those who killed our friends pay.

14

ALL IS LOST

As tensions were rising in Iraq, with the surge in full force, tensions were also rising in my relationship with Rana. I would come home on leave, and she would begin to accuse me of cheating on her or liking one of her friends. It was absurd. I was always on base. How could I be cheating?

One day, these accusations came to a boiling point. I was home flipping through the TV channels when Rana came into the room, giggling as she rushed over to her vanity to grab something. I followed her and saw a few of her friends in the kitchen wearing shorts and tank tops with robes on top. But the robes were open. The women saw me and quickly covered up. Culturally, women do not wear anything revealing in front of men, so this was a big faux pas. I quickly turned and left the room. I didn't think much of it until later when Rana started questioning me.

"You like her, don't you?"

"Who?" I was confused. She mentioned one of her friends I'd seen in the kitchen.

"No," I retorted. "Are you kidding me? I love you!"

"You couldn't stop staring at her!"

This back and forth escalated. It was clear that she had made up her mind that I was unfaithful. She just kept berating me with false accusations as she puffed on her cigarette.

Soon I hit my limit. We started screaming and yelling at each other, our voices echoing off the walls. In a fit of rage, she tried to leave the house. I blocked her in the doorway, getting inches from her face.

"I've never been fucking unfaithful, but have you?" I yelled at the top of my lungs as I punched the wall right next to her.

She was trembling.

I had crossed a line and scared myself in that moment. Immediately, I apologized; I had let rage overtake me. I should have let her walk away and waited until we were both calm to continue the conversation. It had become apparent the cracks in our marriage had grown wider. We were no longer just dealing with her controlling mother or the visa issue. Now, the fissures had escalated into accusations of infidelity. And, that, I just couldn't shake.

★ ★ ★

A few days later, I went out to do some grocery shopping at the nearby open-air market when my cell phone rang. It was my mom, and she was screaming. Her voice was so frantic that I could barely make out what she was saying.

"I just got a call from a friend of your dad's," she said. "Your dad's been in an accident."

Now you need to understand that in Iraqi culture, nobody tells the wife directly when something serious happens to her husband. You tell the oldest male in the family—a son, or the husband's brother—and they convey the news to the rest of the family.

"Okay," I said. "I'll give him a call." I punched in the friend's phone number as I started walking back from the market with my shopping bags.

"Hey, my mom just told me there was an accident," I said. "What's going on?"

"I'm so sorry to tell you," he answered somberly. "Your dad was just shot, and he's dead."

I had my grocery bags in one hand and the phone in the other. As those words hit me, my legs gave out, and I crumbled to my knees right there in the middle of the market. Shock washed over me, numbing everything. Could this really be happening? Tears started streaming down my face, and I began to cry uncontrollably. I struggled to get the words out.

"How'd it happen? Where's his body?" I managed to choke out.

Sensing my distress, my dad's friend took over the conversation, his voice steady but grim as he explained the details. I barely registered what he was saying as I slowly got up, leaving the grocery bags on the ground, walking toward home with my mind reeling.

"He's at the morgue," the friend said. "We had just left work together. Your dad was driving, and I was in the passenger seat. We got on one of the highways, right across from a regular army checkpoint. Suddenly, a car drove by and opened fire. They hit your dad multiple times before speeding off. It was a drive-by."

"Who did it?"

"The Mahdi militia, the JAM."

Dad and his friend worked at the Ministry of Communications office located in a Shia area. The ethnic cleansing was at its peak. Shias were pushing Sunnis out of Baghdad, and my dad was Sunni. Perhaps he was targeted because he was Sunni or maybe even because I was a terp.

"Are you okay?" I asked.

"Yes. I don't even have a scratch on me. The shots came from the driver's side, so I ducked. Unfortunately, your dad took all the rounds."

I hung up the phone. I was in shock. Everything around me was moving at turbo speed, and yet I couldn't even move. I tried to compose myself and figure out how I was going to tell my mom. I waited a few minutes, took a deep breath, and then called her number.

"Yumma, he's gone," I said. I tried to explain what had happened, but I just heard sobbing on the other end of the phone. "I'm sorry."

It was the only thing I could say.

All I wanted to do was go home and be with my mom and my siblings, to console them. But as the oldest son, I knew my responsibility was to collect my dad's body from the morgue and prepare it for burial.

In ordinary times, this would have been hard enough, but trying to arrange everything during the surge was a nightmare. I assured my mom I would figure something out, even though the roads were blocked and our neighborhood was on lockdown. The coalition forces were gearing up to secure the city. Soldiers were going door to door, searching homes for weapons, and they weren't letting anyone in or out of the neighborhood because of suspected Al-Qaeda sympathizers.

I was desperately trying to come up with a plan to reach the morgue when I spotted a US Army Stryker unit parked near a half-built house on a main street. The house was only partly framed, and I figured that it had been turned into a Combat Outpost (COP). I decided to take a gamble and approached one of the soldiers, hoping he wouldn't mistake me for a threat and shoot.

"Hey, I'm a terp. I'm on leave," I whispered between sobs. "I have a family emergency and need to get out of the neighborhood." As I was speaking to him, I saw other soldiers start to come out of the COP with their weapons pointing at me.

"Put your hands up!" they ordered.

I willingly obliged. I tried to stay calm, but I was still crying. So I just kept speaking to them in English, hoping they'd believe me.

"I'm a terp, I have an emergency. I need to get out of the neighborhood. I work with Special Forces. My ID is in my wallet, you can take it. I don't have anything sharp on me. My dad just got killed, I need to go to the morgue."

They took my ID and ushered me into the COP.

There was nothing but rubble inside, plus their gear. They called my unit.

"We've got someone here claiming to be your terp."

I overheard them talking, as they confirmed my ID.

Now that they knew I was one of them, the Staff Sergeant took off his helmet. I felt a brief moment of relief, but the tears kept streaming down my face.

Suddenly, my phone rang. It was my mom.

"Don't go! The JAM is at the morgue waiting for people to collect the bodies of their relatives. It's an ambush," she warned. "Don't go there! Ibrahim will take me to the morgue. He is Shia, and he's closer. He'll get there faster. And I don't want you to get killed, too."

I hung up the phone, enraged that my duty had to be handed over to my uncle. I was at the height of my career as a terp, well established with the Special Forces. I'd been running the streets of Baghdad in full military kit, risking my life for my people, and yet, I couldn't even protect my own family.

I turned to the soldiers, my voice shaking with anger and desperation, and I told them what my mother had just said. They looked at each other, exchanging glances filled with concern and determination.

"What if we went with him?" one of the soldiers suggested. They started discussing splitting their force, taking one of the Strykers and some Humvees to escort me to the morgue.

"That's a different unit's Area Of Operation," one of the soldiers said.

"But he's in our AO, and that's his family," the Staff Sergeant protested. "He's one of our terps."

The soldiers huddled together, voices low and tense, urgently debating how they could intervene. Their mission was simple: protect civilians at all costs. But as they pieced together the tragedy that had struck my family, it became clear they had failed. Their frustration was evident—they wanted to make things right, to undo the damage. But there was nothing they could do. I felt both their empathy and their helplessness.

"Fuck!" the Staff Sergeant yelled, hurling his helmet at the rubble-strewn floor of the COP. He took a knee and looked right at me, shaking his head.

"I am so sorry, Brother."

"Me, too," I nodded in defeat.

It felt like the longest fifteen minutes of my life inside that COP. Now I was very aware that all eyes and ears in the neighborhood were on me. I was sure people had seen me go inside. By now they were probably wondering what was happening.

My focus shifted from trying to get to the morgue to trying to get home without being targeted or having my cover blown. I took a deep breath, glancing at the soldiers who were now looking at me with concern.

"The story everyone knows," I began, keeping my voice low, "is that I work for a cell phone company up north. That's my cover for being away three weeks at a time."

One of the soldiers, a young guy with a serious expression, nodded. I could tell he was trying to gauge the situation.

"But this conversation," I continued, my voice dropping even lower, "has gone on too long. It's compromised my cover. I've got neighbors I don't really trust, and I've reported some of them in the past for suspicious activities."

I leaned over the map lying on the makeshift table, pointing to my house and then to two other houses nearby.

"The second house to the right of mine and the fourth one down the street, they're not just bad—they're connected to Al-Qaeda," I began. "I'm one hundred percent sure of it."

The soldiers exchanged looks. I could see their interest piquing, their expressions hardening. I could feel the weight of their scrutiny and the gravity of the situation sinking in. Their silence was heavy, acknowledging the danger I was in and the risk I'd taken to become a translator.

I took a deep breath, before voicing my plan.

"If you handcuff me, drag me back to my house, search the crap out of my home and make a real scene, I might be okay. It'll make it look like I'm just a civilian."

I saw them considering it, their eyes shifting to the map, to me, then back to each other. One of them, the squad leader, nodded.

"Show us where," he said.

I pointed again, tracing the route from the COP to my house.

"And you should go visit them," I advised, pointing again to the two other houses. "But not right after you take me. Give it some time so it doesn't look too coordinated."

They agreed, and we quickly fell into action. Minutes later, I found myself handcuffed, zip ties digging into my wrists as a foot patrol paraded me down the dusty street. A Humvee rumbled behind us, its engine growling, the soldiers inside scanning the area for any sign of trouble. I could feel eyes watching us from behind curtained windows, peering through doorways, tracking every step we took.

The story we had cooked up was simple: I had tried to leave the neighborhood during the lockdown, raising suspicion. So now the Americans were dragging me back to search my home. It was a good story, believable enough to buy me some time. The soldiers played their parts well, their faces a mask of indifference, hands gripping their rifles as if expecting trouble any second.

As we approached my house, I caught a glimpse of the second house to the right, its windows dark and foreboding. I knew they were watching. My heart pounded in my chest, but I kept my face blank, hoping the performance was enough. The Humvee halted, its tires kicking up a cloud of dust, as the soldiers pushed me forward, showing their authority. My wrists ached from the zip ties, but the pain was a dull whisper compared to the storm inside me. My father had been killed for no reason other than being Sunni. And now, here I was, my own life hanging by a thread. If even one person suspected the truth —that I was working with the Americans—my fate would be sealed.

A surge of adrenaline hit me as the soldiers moved in, rifles raised. They didn't hesitate; with a swift kick, they busted the front door open. The sound echoed through the narrow street, drawing curious faces from every doorway.

Inside, I saw Rana's eyes widen in fear. I caught her gaze and mouthed silently, "It's okay. I'll tell you everything later." Her face was pale, her hands trembling as she watched the soldiers storm in.

People in the neighborhood were already gathering, whispering among themselves, pointing fingers. The soldiers didn't waste time; some blocked the door, maintaining security, while others tore through the house. They yanked open drawers, dumped their contents onto the floor, flipped mattresses, and overturned couches.

Everything that could be turned upside down was turned upside down. Our home was in shambles. Rana was shaking, she stepped closer to me, tears streaming down her face. I wanted to reach out to her, to reassure her, to tell her that this was all a performance, a desperate act to keep us safe. But my hands were bound, and my voice was trapped, strangled by the weight of fear and grief. All I could do was stand beside her and weep.

The soldiers left as abruptly as they'd come, leaving the house in ruins. The door had barely closed behind them when the neighbors started pouring in, their faces a mix of curiosity and concern. Six or seven men crowded around me, voices overlapping.

"Are you okay?" they asked. "What's happening? What's going on?"

I just kept crying, rubbing my sore wrists that the soldiers had cut free before leaving. It was all I could do. I struggled to find words, my throat tight, my mind spinning.

Finally, I managed to blurt out.

"My dad," I started. "My dad was killed. In a Shia area. The JAM killed him. I tried to leave to go to the morgue, but the soldiers—they brought me back. They won't let me leave."

"We're so sorry," they started to murmur. "Is there anything we can do?"

"My mom said the JAM are waiting for any Sunnis to show up at the morgue so they can kill them too. I can't go to be with my family."

Rana's eyes met mine, and I could see that she finally understood the full weight of what was happening. She reached for my hand, squeezing it tightly, as if to anchor us both in the storm.

Our neighbors' voices filled the room, offering words of comfort, assurances that this wouldn't last, that we would be okay. But their words felt hollow, empty promises that only made the ache inside me worse. How could anything ever be okay again?

Suddenly, I realized that instead of having space to grieve I was hosting strangers in my living room.

"Get out of my house!" I told them fiercely between sobs.

As soon as the last of the neighbors slipped out the door, I crumbled to my knees, my legs giving out beneath me. The adrenaline that had kept me upright drained away, leaving me empty and exhausted. Rana sat down beside me, her arms wrapping around my shoulders, pulling me close as we both sobbed. The noise of our grief was all I could hear, drowning out the world outside.

I had always known that being a terp was risky, but until now, I hadn't truly felt the full weight of that risk. The danger had always been there, a shadow lurking at the edges of my thoughts. But now it was front and center, demanding my full attention. Most of my friends had been killed—picked off one by one because they'd made the same choice I had. My grandfather was gone, and now my father, too. Their deaths were like open wounds, each one cutting deeper than the last.

I was lost, drifting in the fog of war, struggling to understand what all of this was for. Was democracy worth it? Was it even possible in a country like mine, where sectarian lines ran so deep? The ideals that once seemed so clear now felt distant and out of reach.

The reality of my situation settled in with crushing clarity. I wouldn't even be able to go to my dad's funeral. It was too dangerous, the threat of being ambushed by the JAM too real. I had been robbed of my chance to say goodbye, to find some semblance of closure. Instead, his death hung over me like a dark cloud and filled me with unanswered questions. Had he been murdered because he was Sunni? Or did my work with the Americans have something to do with it? Guilt gripped me like a vise, squeezing the air from my lungs. I couldn't shake the feeling that I had brought this upon him, that my choices somehow had led to his death.

A dark voice in the back of my mind whispered of revenge. It was the only path that seemed to make any sense. The desire for justice burned like fire in my chest, becoming all consuming. I was spiraling, engulfed by the need for righteous retribution.

15

I'M THE MONSTER I FEARED

I cut my leave short and returned to base after spending a few days trying to restore as much order to the house as possible. But I couldn't stand being home anymore. I felt helpless, and all I wanted to do was get revenge. I was fueled by hatred and had a mission to avenge the death of my father. At this point my moral compass had been completely gutted. I was out for blood.

When I arrived back on base, the 19th Group was in rotation, and Bob was my new Delta Company advisor. I felt numb, the events of the past few days a blur in my mind. Grief and anger swirled together into a toxic mix.

I sat in my room alone, brooding, when I heard a knock at the door.

"Come in!" I called out.

I turned my head to see Bob in the doorway. He made his way over to me, concern etched into his features. I knew he must have heard the news.

"What's going on?" he asked quietly, his voice low enough that it wouldn't carry to the hallway.

I swallowed hard, trying to find the words, but then they just tumbled out.

"My dad was killed," I said, my voice breaking. "The Mahdi Militia got him. He was just... in the wrong place."

Bob didn't say anything, patiently waiting for me to continue. I took a shaky breath, the anger rising to the surface.

"I need a pistol," I said. "One of those unregistered ones we find during raids. I know we keep some of them for training purposes. I want one to take outside the wire."

Bob's eyes narrowed slightly, and I could see the wheels turning in his head. He knew why I wanted that pistol. He knew exactly what I was planning to do with it. I met his gaze, my expression hardening, all the rage and pain spilling over.

"I just want to kill some fucking Madhi Militia," I said, my voice cold and flat. "I want to be the triggerman."

For a long moment, Bob just looked at me, his eyes searching mine. He didn't need to ask if I was serious; he could see it all over my face. He knew I was a ticking time bomb, ready to explode. He also knew that if he gave me what I was asking for, he might be sending me to my death —or worse, into a situation where I'd take others with me.

"Listen," he said finally, his voice steady and calm. I felt like he was trying to talk me off the ledge. "I get it. I know you're hurting, and I know you're angry. But giving you a pistol isn't the answer. It's only going to make things worse."

I shook my head, frustration reaching its peak.

"I don't care, Bob," I replied. "I need this. I need to make them pay."

Bob sighed, a long, weary sigh. He reached out and put a hand on my shoulder. He was firm but not unkind.

"You're not thinking straight, Waleed," he said. "Right now, you're a liability. I can't let you go out there like this. I'm grounding you. You need time. No more missions until you get your head straight."

His words hit me like a punch to the gut.

"You're grounding me?" I repeated in disbelief. "You're taking me off missions?"

"It's for your own good," Bob said firmly. "And for the good of everyone else. You can hate me all you want, but this isn't up for debate."

I stared at him, feeling the walls close in around me. The need for revenge was eating me up, but Bob's eyes were unyielding. He wasn't going to budge. He turned away and walked out of the room.

The anger inside me raged, and I felt like I might explode from the sheer intensity of it. Deep down, I knew Bob was right; I knew I wasn't thinking straight, that I was a danger to myself and everyone around me. But that understanding did nothing to extinguish the fire raging in my chest. It didn't ease the raw, desperate need for justice coursing through my veins.

I was grounded for a few weeks, but it felt like an eternity. Every day stretched out endlessly, and I hated every minute of it. Being stuck on base, watching others come and go, was torture. I felt trapped, like a caged animal, with nowhere to channel all the rage that was roaring inside me. I could only sit there, stewing in my own thoughts, wondering if the terp who replaced me was okay. Every time the guys went out on a mission, I felt the weight of guilt pressing down on me. I didn't want anyone else to get killed because of me.

I knew I had to snap out of it. If I couldn't control my anger, I needed to at least fake it, to convince Bob and the others that I was ready to go back out. Sitting around doing nothing wasn't an option. I had to get back on missions. I had to do something.

A few weeks later, Bob came back to my room. He told me he had just finished briefing the Delta Company Commander about a new mission. As he turned to leave, he paused at my door and looked me over.

"You ready to get back out there?" he asked, his voice measured.

"Yes, I am good to go," I replied, my answer coming out quickly and firmly. I tried to project calm confidence to mask the turmoil still churning inside me.

Bob studied me for a moment, then nodded. "All right, you're back in."

The next few missions were the usual routine: targeting Al-Qaeda operatives or the JAM. I was back in the field, back where I felt like I could do something, like I could make a difference. But every time we went out, I found myself being more generous with my ammo, my finger itching on the trigger. Each shot I fired, I pictured it hitting the bastards who had killed my father.

A few weeks later, I finally got my chance. It was the 5th Group's rotation, and we had a mission to hit a Mahdi Militia office—one of their main spots for operational planning. The briefing was straightforward: we were to go in hard and fast. We prepped for the mission, running through rehearsals in the afternoon. Then we geared up, grabbing weapons and ammo from the arms room. The Humvees were lined up outside Delta Company's HQ. It was go time.

We drove in blackout mode as usual, the night swallowing us whole. But when we approached the target area, the streetlights were on. The power grid was up. Now we were fearful someone might see us and tip off the target. We flicked on our headlights, and gunned the engines, speeding through the narrow streets until we reached the target house.

The raid was swift. We dismounted the vehicles in seconds, forming a line against the front gate. With a loud crash, we breached it, and the assault team surged forward. I moved with them, heart pounding, eyes scanning for threats. To the left, there was a narrow walkway.

Suddenly, a shout pierced the air. "Man on the roof!"

I bolted toward the walkway, weapon raised, eyes searching the rooftops. Then I saw him—a man in a dark tracksuit, trying to leap to the roof of the neighboring house. Without hesitating, I aimed and squeezed the trigger, firing off a couple of quick shots. The man stumbled, but I couldn't see if I had hit him or not.

Inside the house, chaos erupted. The assault team was met with a hail of AK-47 fire. One of our guys went down, hit in the calf. I sprinted into the house, finding three detainees already on the floor, handcuffed and under guard. Beyond that room was another room, and from the sound of it, a terrorist armed with an AK-47 was firing at anything that moved. Nassir, an ISOF soldier, aimed a SAW machine gun through the doorframe, pointing it towards that room. I quickly translated for SSG Gwen, a USSF medic standing nearby, what was happening.

He looked at Nassir and said, "Let's get this son of a bitch."

Nassir took the lead, moving into the first-man position. I fell in behind him as the second man, with Gwen as the third man of the stack. Nassir pulled the pin on a flashbang and lobbed it into the room. A deafening bang and blinding flash followed. Then we were moving—Nassir fired the first shot as he entered the room, and I was right behind him, squeezing off a few rounds. Gwen came in next, firing to finish the job. The terrorist didn't stand a chance. He went down under a hail of bullets, his body riddled with holes.

With the immediate threat neutralized, we began the Sensitive Site Exploitation (SSE). I started tearing through drawers, flipping through papers, looking for anything that could be intel. Across the room, Nassir stood over the terrorist's body, ensuring he was down for good. The US intel team rushed in, quickly assessing the scene. As I rifled through the debris, I noticed a notebook lying on a credenza. I picked it up and started to flip through the pages. My face must have given something away because one of the intel guys immediately zeroed in on me.

"Waleed, what is it? What did you find?"

"Motherfucker! This is a kill list." I showed them the notebook. "Look at all the names. And look! Some of the names also have dates of when they were killed. There were dates as recent as two days ago!" I shouted.

Something inside of me snapped. Could this have been the terrorist that killed my father? I walked over to the dead terrorist's body and stomped his face.

"FUCK YOU!" I screamed over and over again.

I finally composed myself enough to exit the room. In the hallway, one of the detainees was sitting on the floor screaming.

"What did you do to my brother?" he wailed. "Where is my brother?"

I took a knee in front of him and pulled down my brown gaiter to reveal my face. I wanted him to see me. I locked eyes with him.

"I shot up your fucking brother," I answered him "He is as dead as all the people you guys have killed. Keep screaming, and I will put a bullet in your fucking head, too."

I pushed his face to the floor as hard as I could and left him there.

On the ride back, my stomach churned with a mix of nausea and regret. I never imagined I would cross that line. I had become the very monster I was trying to eliminate. Revenge had become my compass, and I no longer recognized myself. The tightrope, on which I had been delicately balancing since the war began, snapped that day.

As soon as we got back to base, I just wanted a drink. Long gone were the days when alcohol was a companion only seen at college parties. Now alcohol had become my mistress, comforting me before I laid my head to rest—almost every night. This transition had been a gradual process. First, it was about having a good time with the other soldiers, then it was about finding a way to relax. And now, it was about staying numb. I had no idea how to cope with the reality that I'd taken a life so ruthlessly, for the first time, at close range. I had seen his face—so different from the distant targets in past missions. I had craved this moment, but killing him hadn't brought relief. It only made me feel worse. Now, I just didn't want to feel anything at all.

So I chose my favorite mistress that night: Bells Scotch, a strong whisky. I plummeted into an abyss of darkness and despair, haunted by the face of the man I had shot. The echoes of my own choices reverberated through my mind, tormenting me through the sleepless parts of the night and then again in my dreams.

★ ★ ★

My Islamic worldview began to unravel. Every time I went out on a mission I saw devastation and destruction by fighters who claimed to be fighting a holy war. Yet nothing they did was holy. Seeing the worst in humanity every day pushed me to believe there was no God. How could God allow these horrible things to happen to innocent people? Why did murder, kidnapping, rape, sectarian violence, war, car bombs, and beheading exist? My country was in shambles and these horrible acts were a daily occurrence. Anger consumed me, and alcohol numbed me. I was no longer a Muslim, and there was no longer a God, as far as I was concerned.

★ ★ ★

The following days were miserable as the new team arrived. The new US team leader focused closely on the translators assigned to HHC, prompting COL Ali to transfer me from Delta to Alpha. This change introduced me to the new USSF Alpha advisor, SFC Ryan. Unlike Wil, Ryan wasn't interested in building a relationship. His indifference was unmistakable, making an already challenging situation even more isolating.

Alpha was, from my perspective, the bottom of the barrel in terms of lethality, quality of training, and quality of troops. It was a graveyard, both professionally and in terms of risking death.

About three weeks later, Alpha was tasked with returning to Al-Shu'ala to capture a terrorist. This time, we left the Humvees behind and rolled in with Mine-Resistant

Ambush Protected vehicles (MRAPs), heading straight into the neighborhood where I used to live. We parked on a side street to keep the large trucks hidden and left a group of Iraqi commandos to secure the vehicles.

We then moved out on foot, crossing the main street—a two-lane road in each direction with a six-foot-wide median. Our EOD team went ahead to sweep for IEDs, ensuring the area was clear and safe for us to move. Once they gave the all-clear, we sent a team to secure the crossing, made our way to the other side of the street, and continued our approach.

Once you kick in that first door and make noise, you need to get in and out quickly. We stormed the house and broke in, the women inside started to scream. The chaos was immediate. We identified the men, separated them, and tried to Positively Identify (PID) our target. We had to move fast. One of the USSF guys had the document with the target details and confirmed we'd found the right guy. We handcuffed him and began to walk him out of the house.

There's a certain way to handle a detainee. Once handcuffed behind his back, you stick your hand up from below the cuffs, between his hands and his back, and grab him by the back of the collar. Then you force his hands up high so he has to bend at his waist as you walk. I was holding him that way, controlling him while still holding my weapon. As soon as we stepped out of the house, he tried to scream, making a lot of noise and calling for people.

I immediately punched him in the ribs, knocking the air out of him.

"If you're not quiet, I'm going to shoot you right here and dump you in the street," I snarled, my voice low and menacing.

I couldn't have followed through, but the threat worked. He fell silent, fear in his eyes. We had to get that high-value target out of the neighborhood quickly.

We loaded everyone back into the trucks, the tension was palpable as we began to drive through the streets. Every shadow, every corner, felt like a potential threat. The memories of my old neighborhood, now a battleground, only added to the surreal and unsettling atmosphere. The detainee was crying as we raced out of the neighborhood, knowing that every second we lingered here increased the risk of an ambush.

We got to the base, turned the detainee in, and then started our AAR. The tension of the mission lingered in the air, a mix of exhaustion and adrenaline putting everyone a bit on edge. We gathered in the briefing room, the fluorescent lights casting a harsh glow on our tired faces.

Ryan took charge.

"Congratulations guys," he began. "You just captured the son of a bitch who blew Delta Company up."

I turned to him, startled. In a flash, I was taken back to that moment when Bassim had died in the seat that should have been mine.

"Wait," I interjected. "What? Why didn't you tell me?"

"Shut the fuck up and translate," he snapped, oblivious to my past. He had no idea that I had been on that Delta mission those few months ago, that I had lost several friends and had nearly gotten myself killed.

I went ahead and translated. The Alpha Company Captain asked the same question.

"Why didn't you tell us?"

I translated, and the advisor snapped at me again.

"I told you," he said, annoyed. "Just translate."

"That's not me," I told him. "That's what he's asking. He's asking you the same question."

Ryan turned to one of the more politically attuned team members.

"We know you guys take this personally," he said. "So we didn't tell you."

The Alpha SF advisors had known exactly who we were going after, and they also knew why they didn't share that information until after the mission was complete. They wanted the terrorist alive. If we had known we were going after the terrorist who had detonated the EFP bomb on Delta Company, we would have likely killed him on sight.

★ ★ ★

I craved a relationship built on trust with my new team of advisors, but they were only mission-focused, with no interest in building rapport with their Iraqi teammates. It pissed me off and set me even more on edge than I already was. Every mission drained a bit more of my spirit. I was constantly caught between the loyalty to my country and the frustration of being sidelined by the very people I was risking my life to help. The war had taken so much from me already—friends, family, any sense of normalcy. And now, it was taking my hope, too.

I found myself constantly questioning the purpose of it all. Was it worth it? Was I making any difference at all? Or was I just another casualty in a conflict that seemed endless and unresolvable? The weight of these questions bore down on me. After six years of war, the urge to walk away grew stronger with each passing day.

On my next leave cycle, I snuck over to my mom's house to check on her. The shock of losing my father had taken a heavy toll on her, and I could see it in her eyes. It was late afternoon when we sat in the dimly lit living room, the air permeated with the aroma of chai and unspoken regrets.

"Yumma, how are you?"

"I'm okay," she replied, trying to maintain her poise, fighting back tears.

"Yumma, how are you really?"

She began to cry.

"I'm devastated, and I have no idea how I can afford to take care of your siblings."

I scooted closer to her and reached for her hand.

"That is something I wanted to talk to you about," I said, hoping to give her any comfort I could. "Kris, my medic friend, told me that he would be my sponsor and help me get a visa to go to the United States. It will take some time with all the paperwork, but I already started filling it out."

My mom began to cry harder.

"I can't lose you, too."

"But, Yumma, if I stay here, I will get killed. At the rate I'm going, I won't make it to thirty. The risk is constant. At least, if I am in America, I can send money back to help take care of you all, and I can apply for your visas."

She was quiet. She knew I was right. She knew she would be losing me but at least I wouldn't be dead.

After a few minutes she shook her head and said, "Inshallah."

If God wills it.

My mom had already endured so much heartache and loss, and the last thing I wanted was to add to her pain. But I also knew how resilient she was—one of the most important lessons she taught me was to keep going, no matter what.

I knew she'd be okay. But now, I faced a much tougher challenge: convincing Rana that moving to America was a good idea. I knew that would be far more difficult than convincing my mom.

16

REGRETS AND REMORSE

Qutt had already gotten his Special Immigration Visa and was living in America. I, however, was still waiting on my paperwork and for Rana to agree. We would get into fights about moving to the US almost everyday.

"I don't want to go with you!" she would cry. "I want to stay with my family."

I'd try to convince her by bringing up our traditions.

"I'm your family now," I'd say. "You need to go with your husband."

She wasn't hearing it.

At that point, I pretty much knew where I stood. The last straw came the day that I saw one of our male neighbors walking down the street, using the iPod music player I'd given to Rana as a gift.

I confronted him.

"I borrowed it from your wife," he said.

I went home and asked Rana.

"Where's your iPod? We should listen to some music."

"Oh," she said, "I left it at my parents' home."

Was she cheating on me? The thought gnawed at me. I was only home one week a month, leaving her plenty of opportunities. Her constant accusations about infidelity began to feel like, maybe, she was trying to make me feel guilty for something she was doing herself.

Desperate for clarity, I reached out to Rana's aunt and asked if we could meet for lunch. I hoped for some marital advice, but what she told me was far more than I had bargained for.

"When your wife miscarried," she began gently, "it might have been because she was taking birth control pills."

I stared at her, stunned. We had never agreed to postpone having a family. The idea that she could do this without telling me felt like a betrayal.

"And there's something else," she continued, her voice tinged with reluctance. "Your wife didn't break off her first engagement on her own. Her mother did it when she saw that you were a better prospect, financially."

I felt like I'd just gotten the wind knocked out of me. All this time, I had believed the narrative she and her family had spun. Now, it seemed our entire relationship was built on lies and manipulations. How could I trust her?

I left lunch feeling confused and betrayed. How could I repair our relationship when it was built on such a shaky foundation? Trust, once broken, is a hard thing to mend. As I mulled over the revelations on my way back to our place, the weight of the truth settled heavily on my shoulders. She was a liar. And I didn't know if our marriage could survive this.

But before I could even try to repair our marriage, Rana's mother decided she wanted her daughter to divorce me. A few days later, she filed for the divorce. In Iraqi culture, if a husband files for divorce, the bride's family receives a payment, but if the wife divorces him, she owes him nothing. So, even though I had provided our home and everything in it, she wouldn't owe me a single thing.

The power struggle I had from the beginning with my mother-in-law continued even after receiving the divorce papers. In the home Rana and I still shared, I had a lot of photos from my military service at Camp Slayer. We didn't have digital cameras back then, so I'd shoot with a film camera and take the rolls to a local photographer's storefront studio to have them printed. The US soldiers did that, too. A lot of people participated in the local economy without formal military contracts.

I can see now, looking back, that having those kinds of photos printed by a nonmilitary shop was not very safe. But I'd known the guy well enough to trust him. Eventually, as things heated up, he told me to stop.

"I don't want to print these pictures anymore," he had said. "Don't bring them to me."

The one I really shouldn't have trusted was my mother-in-law. She came to me one day brandishing a handful of my photos.

"If you don't go through with the divorce, I'll give all your pictures to the guys here," she said. "I know plenty of them."

The guys she was talking about were the Sunni extremists and Al-Qaeda loyalists in the area. At that point, it became entirely clear that she wasn't just going to undermine our marriage, she'd readily have me killed to put an end to it.

There was nothing left to fight for. The road ahead was uncertain, but one thing was clear: there was no going back. I signed the papers.

★ ★ ★

A few days later, I took one of my college friends and his wife to dinner at Al-Sa'a, which means 'The Clock.' It was a fancy restaurant with round wooden tables and leather seats, located in the upscale Al-Mansour neighborhood. The atmosphere was elegant, with marble floors, granite walls, and glittering chandeliers adding to the luxurious feel of the place.

We were enjoying our delicious shawarma when my friend suddenly leaned over to me and lowered his voice to a whisper.

"Waleed," he said quietly. "Look at me. Whatever you do, don't look behind you."

His statement was odd, and the way he said it immediately made me feel suspicious about whatever was happening behind my back. My heart began to race.

"What's going on man?" I shot back. "Tell me!"

Slowly, I started reaching for my gun.

"If you want to do something, I'll do it with you. Your ex and her mother just walked into the restaurant."

"Oh." I breathed a sigh of relief. "So what? I don't care."

I'd thought there was some sort of a threat behind me.

"They're not alone," my friend replied. "There's a guy with them. Does your ex have a brother?"

"No, she does not. You know that," I snapped back at him.

But as soon as the words left my mouth, a realization hit me like a freight train. I froze, feeling the blood drain from my face. All the time she was leaving to be with her parents, the pills, the iPod she'd given away, the accusations of infidelity, the coldness in her eyes—it all suddenly made sense. There had been someone else in the picture the entire time.

A tidal wave of betrayal washed over me, leaving me hollow and disoriented. My heart ached as the truth settled in, cutting deeper than any wound I'd ever felt. I turned my head and saw her sit down. They had spotted us, too. And out of all the tables in the restaurant, they had purposely chosen the one right next to ours.

It felt like a cruel joke. I watched as she laughed, acting like she was unaware of my presence, her eyes lighting up in a way I hadn't seen in a long time. It felt like a knife being twisted in my gut—she was happy, and I had been the fool.

Betrayal suddenly turned to rage. I couldn't just sit there and watch them. The sight of her with him, so casual, so shameless. And her mother supervising the whole thing. It made my blood boil. I leaned over to my friend and his wife.

"Let's get out of here," I said.

Without a second thought, I stood up, pulling a thick wad of cash from my pocket. I ostentatiously took out far more than was needed to cover dinner and the tip and dropped it onto the table, letting the bills scatter as they fell. The gesture wasn't about the money—it was about making a point.

I turned to my ex-wife, catching her off guard.

"The money I spent on you is just money," I said, my voice cold and steady. "Money comes and goes, but only dirt can fill the greedy."

With that, I walked out, leaving her stunned and speechless. As I stepped outside, the cool night air hit me, and for the first time in a long while, I felt a sense of closure. Now, at least, I knew the truth.

★ ★ ★

Rana had the audacity to call me two days later.

"We have a week left on the lease," she said matter of factly. "I'm trying to sell the furniture, and I need help."

"Why are you calling me?" I asked brusquely. "There's no 'we' anymore. And I gave the furniture to you."

"I don't have anyone to help me move it," she whined. "Can you buy it from me? Just take it and give me cash for it."

Seriously? I had already bought that furniture once. Then, I'd given it to her in the divorce. And now she wanted me to buy it from her?

"I don't have enough cash," I replied. "Let me get back to you in a few days."

I let a few days go by to kill the time she had left on the rental contract. She called me again to see if I had come up with a solution.

"You really thought that I would give you money for what I bought you?" I laughed and hung up on her.

I was done. That was my last conversation with her ever.

I spent a few months revising the immigration paperwork to remove Rana's name from the forms. That change delayed my visa by almost nine months. It also meant, for some bureaucratic reason, that since Rana had been on the initial filing, I couldn't revise it to bring my mom and siblings to the US with me instead. If I'd been single when I started the process, I could have brought them with me. Instead, I now faced the reality of moving halfway around the world without them. The regret over my marriage weighed heavily on me, but what cut even deeper was knowing that my poor decision would once again separate me from my family. This time, though, there would be an ocean between.

A strange shift took place during the next few months, which ended up being my last months as a terp. Not only did I not connect well with the team, but my outlook during missions changed. When I had first started in this line of work, at the beginning of the war, I had made peace with the idea that I might get killed. Accepting that helped me do my job better. Now, with the end of my service in sight, I had not only survived—I had something to look forward to. For the first time, I could see a future, a life I might actually enjoy. Oddly, that made me fearful on missions in a way I never had been before.

Now, I was always thinking, "I don't want to get blown up; I'm getting ready to leave." It was an unexpected, uncomfortable side effect of hope.

The day I received my visa was surreal. For so long, I had imagined this moment, counting down the days until I could hold that small, powerful piece of paper in my hands. Now it was here, and I felt a pang of heartbreak.

I walked into the living room, holding my visa in my hand. My family was gathered there, waiting, their eyes red-rimmed and glistening with tears. I felt both happiness and deep sadness at the same time, torn between two worlds.

My mother stood at the center of the room, her shoulders hunched over as if burdened by an unbearable weight. When she looked at me, her eyes were filled with tears. And I felt my heart shatter. She had been my rock through everything, despite our recent past. And now her silent, tear-streaked face was pleading with me not to go. Her lips quivered, but she said nothing. She didn't have to; her grief spoke louder than words ever could.

I stepped toward her, but my feet felt like they were moving through quicksand.

"Yumma," I whispered, my voice cracking. I wanted to tell her that everything would be okay, that this was for the best. But the words stayed stuck in my throat.

She reached out, her hand trembling as she touched my cheek.

"Why must you go?" her voice was a whisper, broken and raw. Her tears flowed freely now, her sobs quiet but filled with so much pain, as if she was mourning my death.

I swallowed hard, trying to hold back my own tears, but it was impossible.

"I have to, Yumma," I said softly. "It's the only way."

She simply nodded.

★ ★ ★

Leaving my family behind was gut-wrenching, like a part of my soul was being ripped away. The visa in my hand was supposed to represent freedom, but all I could feel was the pain of separation that was as real and tangible as the tears on my mother's cheeks.

Alongside the sorrow, there was a conflicting sense of guilt. It felt like I was betraying the comrades with whom I had fought. Like I was not honoring the bonds we had forged in the crucible of war, where we had shared blood, sweat, and tears. Walking away from the battlefield without losing a limb or my life didn't seem right. And it left me grappling with a complex mix of emotions.

Most of my friends had fallen, and their memories remained a haunting echo in my mind. Yet here I was being offered a new life, a second chance. This visa represented more than just a ticket out of a war zone; it was a lifeline, an opportunity to start anew, to carve out a future free from the terrors of conflict.

Preparing to leave was bittersweet. I had to get rid of a lot of belongings. I would be traveling as a civilian with just one suitcase. I wished I could have kept my uniforms, especially the desert camo that had been tailored just for me. There were so many memories wrapped up in these for me. The base tailor had made an innovation in our gear that would later be copied throughout US combat forces; he had moved the uniform pockets from the chest to the sleeves so we could access them while wearing body armor. I had to let go of my really nice Blackhawk vest and the refrigerator I'd gotten for my room. Then there were all the logistics. I terminated my internet service, disconnected all the cables, and sold my gear: the computers, the tower, the satellite dish.

The night before I left, I stayed at my family's home. My youngest sister Amel, who was six or seven at the time, asked to sleep in my bed with me. She snuggled up and held me close. I couldn't sleep all night, thinking about what life in America would be like. As the sun came up, I quietly slipped out the bed making sure not to wake Amel. I thought it would be harder on her if I woke her up to say goodbye, so I just disappeared. Now, when I look back, I think about how messed up that was. My heart literally breaks knowing that one of the only father figures she had just disappeared out of her life without as much as a goodbye. This is one of my biggest regrets.

A good friend from college had offered to drive me and my mom to the airport so she could see me off. As we approached the airport's main gate and checkpoint, my mother held me tight, weeping. She started kissing me all over my face.

We all got out of the car and waited for another friend who'd offered to escort me through the checkpoint. He still had a badge, so it would go much faster that way.

As we stood in the gravel lot waiting, watching the bustle of vehicles at the checkpoint, I tried to console her.

"It's gonna be okay," I said. "I will apply for your visa as soon as I get there, and we will be reunited. I survived this whole time. I can't just sit here and wait for the JAM to get me. I have to go for everybody's sake."

She clung to me.

"I'm just going to go and check things out," I continued. "Give it six months. You'll be joining me, or I'll come back."

Of course, I knew better. But at that point, I'd become used to lying to her about everything I did because I didn't want her to be frightened.

The line of contractors, soldiers, and other travelers waiting to get through the checkpoint and into the airport ground was growing by the minute. The atmosphere was tense and buzzing with anticipation.

Eventually, my Iraqi commando buddy who'd offered to escort me, pulled up in his truck, motioning for me to get in. I picked up my one suitcase, which contained my entire life's possessions, and turned to my old college friend. I hugged him tightly. Then I turned to my mother for one last embrace.

As I pulled her close, I whispered into her ear, "Yumma, we will see each other soon. I promise."

The weight of her sorrow and my unspoken fears intertwined, creating a moment that felt both eternal and fleeting, a desperate grasp at a promise we weren't sure we could keep. Would we really see each other again? This question hung heavy in the air. I honestly didn't know what the future would hold, but we both clung to the glimmer of hope we had, refusing to let it go. In that embrace, we shared a resolve to believe that somehow, against all odds, we would be reunited.

I wiped away my tears, and let go of my mom. With a heavy heart, I turned and got into the truck and passed through the security checkpoint, glancing back one last time. As her figure grew smaller in the distance, the weight of our shared promise remained, pushing me forward into an uncertain future.

PART 3

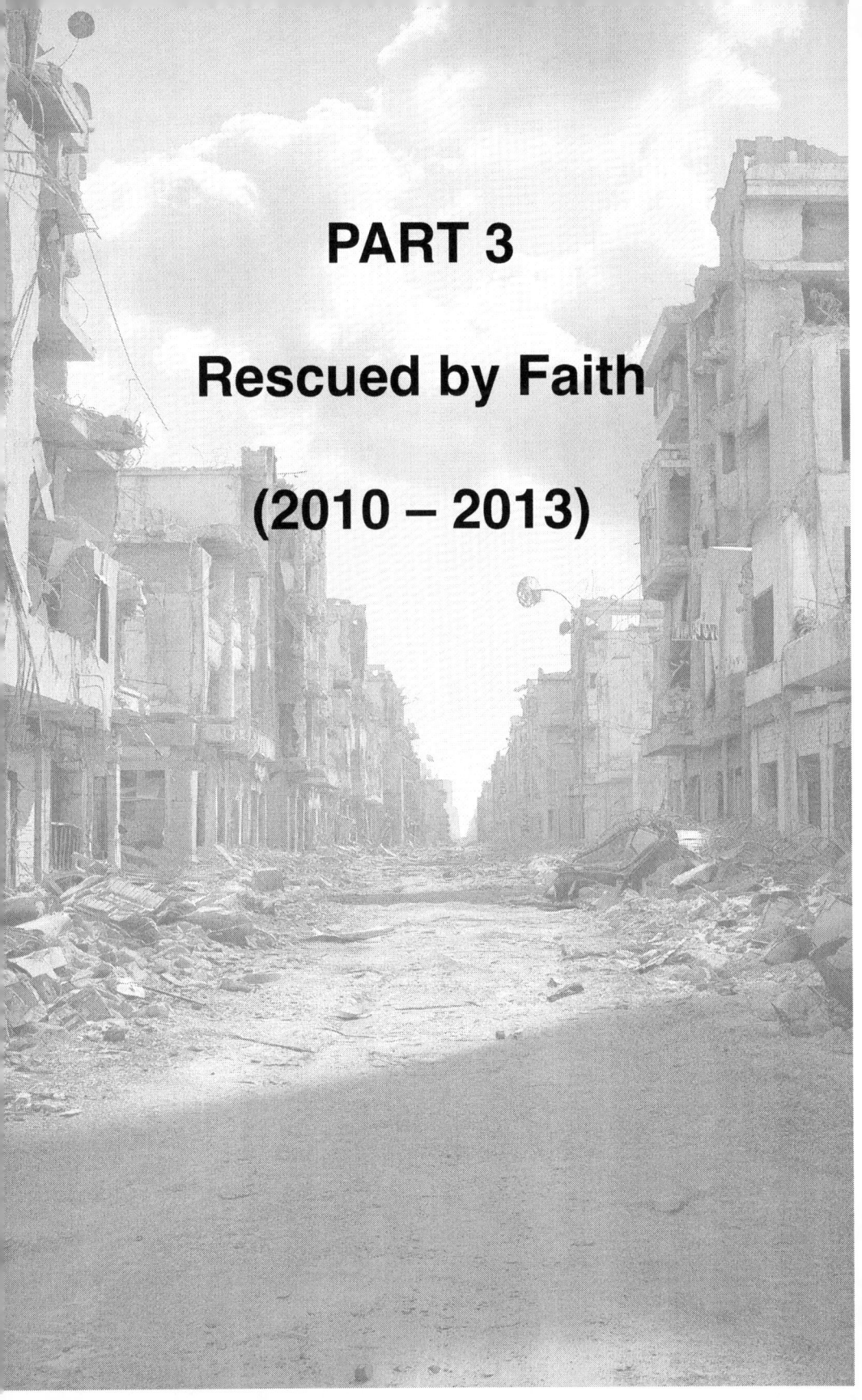

Rescued by Faith

(2010 – 2013)

17

ANOTHER CHANCE AT FREEDOM

My Iraqi Army escort got my suitcase out of the truck for me and bid me farewell. "All the best, buddy. Be safe, and keep in touch." And I walked into the terminal building at Baghdad International Airport.

I boarded my plane at 10:43 a.m. along with about forty other civilians. It was my first time on a civilian aircraft. I buckled my seatbelt and looked out the window. As we lifted off from the runway, I took in the Baghdad skyline one last time. I closed my eyes and sighed with relief. I had made it out alive. I determined right then and there not to look back but to focus on my future. I had my entire life ahead of me. I was only twenty-nine.

As the time in flight went by, my anticipation grew. I was getting closer to the America I had seen in the movies. The place my Special Forces friends came from and would talk about. The place that held the American dream. Freedom was at my fingertips.

My first stop was in Amman, Jordan. While transferring between flights, I approached passport control and handed my passport to the immigration officer. As he examined it, he began to question me, his gaze shifting between me and the document.

"Your name?" he began.

"Waleed Hamza."

"Where are you coming from?"

"Baghdad."

The officer pressed on. "Where are you going?"

"To America. See my visa."

The immigration officer flipped through the pages of my passport, his eyes narrowing as he scanned his computer screen. His brow furrowed with concern, and after a moment, he motioned for a security guard.

The guard approached the booth with a stern expression.

"Come with me," he barked, as the officer handed him my passport.

At this point, I was sweating. Was I being stopped because there was some mix-up or because I was a translator? One thing my life had taught me so far was that you couldn't trust anyone, and things weren't always what they seemed. My thoughts began to race as the guard led me to a small room and asked me to sit down. He left, and the minutes dragged on, feeling like hours.

Finally, he returned and apologized for the inconvenience. Apparently, there was a wanted terrorist with my same name. Quickly, I gathered my things and ran to the gate to board my next plane. Next stop: Chicago O'Hare International Airport.

As the plane approached Chicago, I started getting excited. The sky was clear. And from the airplane window I could actually see cars on the roads below. I was used to landing in the desert where dust always clouded the air. And I'd never seen so much greenery. It was beautiful. We got winter rains in Baghdad, but it never looked like this. Everything here seemed so lush.

Once inside the airport terminal, I saw people wearing vests marked 'World Relief,' the non-profit that had sponsored my case and had helped me with the paperwork. They were now here to help me get through US Customs and then find my connecting flight to Raleigh-Durham International Airport which would be my final destination.

The original plan had been for me to stay with a US Vet in Colorado, but my friend Qutt had urged me to come to Raleigh, North Carolina, instead.

"It's a better place," he said. "It's quieter. Also, Pete and Dan, the advisors we worked with, are here, and they'd be able to help you out."

Qutt had already been in North Carolina for a while and was then working as an Op4 at Fort Bragg. In that job, he was helping train Special Forces teams using role playing scenarios that included translators. Qutt was living with his family in a townhouse apartment complex in Raleigh. His home was near lots of stores, and it would be a convenient temporary location for me, especially since I didn't have a car.

I landed in Raleigh. I'd finally reached my new home in the West. Pete and Qutt had my flight information, and the plan was for them to pick me up from the airport. I deplaned, found my way to the baggage claim, and waited there with my single suitcase. Finally, I spotted their familiar faces in the crowd.

"Yay! Whassup?" they exclaimed.

I hugged them.

The three of us loaded into Pete's Mini Cooper. I asked Qutt if I could use his phone to call my mom.

"Yumma, I've arrived," I told her. "Everything's okay. I wish you could be here. I love you."

As soon as I hung up the phone, Pete launched right into his version of basic training for a newcomer, a short rundown on what not to do in America.

"You're probably wondering what you're going to do here," he started. "Well, don't do anything stupid. Don't do drugs. Avoid places that say 'Checks cashed here.' Just go and get yourself situated." Unfortunately, he didn't give me a heads-up about dating in America.

On my second day in America, Dan swung by to take me out. He wanted to show me around Durham, where he lived. We went bar hopping, moving from one place to another. I met a few women, but I got the vibe that most of them didn't dig Middle Eastern men. It felt a little disheartening. That was, until we reached the last bar of the night.

There, I met a kind woman who actually took an interest in me. We struck up a conversation, and before long, we exchanged numbers. We even made plans to meet at a coffee shop the next day. As we left the bar, I turned to Dan and grinned.

"I got her number," I told him.

Dan smiled back, giving me a nudge.

"Nice." he replied. "Second day in America, and you're already getting women."

It felt strange to me. Back in Iraq, you didn't just go up and talk to women like that. This freedom was new and thrilling, like a whole world of possibilities was opening up right before my eyes.

The next day, eager to keep our plans, I called a taxi to drive me from Raleigh to Durham to meet her. As I sat in the backseat, I watched the meter climb steadily. Twenty dollars. Thirty dollars. My heart started to race. How much was this going to cost me? Fifty dollars. Eighty dollars? The numbers kept rising, and I started to sweat. Was this guy taking the long way just to rack up the fare? By the time we arrived, the meter read $120. I stared at it in disbelief.

"At this rate, I'll be broke in a few weeks," I thought.

I walked a few blocks to the coffee shop and waited for the woman to arrive. In the military if you are not fifteen minutes early, you're late. I was early. She finally arrived, and we ordered a cup of coffee and sat down. She asked me about my story and how I got to the United States. She seemed really interested in me and wanted to know more. I was very happy to share my story and started to ask her about herself. I was just taking it all in, mesmerized by the conversation, when I leaned forward to give her a compliment.

"You're a really big woman," I told her.

She looked at me with disdain. "What did you say?"

I repeated myself. She abruptly put her coffee cup down and walked away.

I was so confused. What just happened? I sat in the coffee shop alone for a few minutes trying to replay the events in my head. I couldn't work it out. I decided it was best to call Dan. I'd ask him. Plus, I couldn't afford a taxi home.

"Hey, dude. I'm in Durham at a coffee shop," I said when he answered the phone. "I was meeting that girl from the bar."

"Cool. How did it go?"

"Not so good, and now I'm stuck here because the taxi cost me $120. Could you come pick me up and take me back to Qutt's house?"

"Of course. You should have told me you were coming to Durham, dude. I would have picked you up. You don't need to waste your money on taxis."

Dan came to the coffee shop, and I told him in more detail what had happened with the woman.

"Wait, you called her big?" he laughed. "Oh yeah. That's my bad. I should have told you. You never call women big or fat in America, that is the worst thing you can do. It's not a compliment; it's an insult."

We both laughed. It all made sense now. But I also felt awful that I had unknowingly insulted that poor woman. And I realized that now I was going to be the one who would need a terp to translate American culture for me.

Over the next two days, Qutt helped me with things I had never thought about. He helped me get a bank account because, as he explained, everyone uses a debit card here. He helped me get a US cell phone and an apartment nearby. He even introduced me to another Iraqi who lived in that complex.

Even though I'd brought my life savings with me, and even though World Relief had given me a small loan to cover my airfare, I still had to borrow some money from Pete to pay the deposit for the apartment. Pete's wife took me shopping to get some pots and pans for my apartment. The prices were shocking to me.

"I can't afford this," I told her, stunned by the cost of even the most basic items.

It quickly became clear that my life savings wouldn't last long in the US. It was a harsh wake-up call, and I realized the road ahead would be full of challenges as I struggled to find my footing in this new country.

Unpacking took me all of five minutes. I unloaded the few clothes from my one suitcase and put them on the floor. Then I took out the few other items I'd brought, and closed my suitcase.

My apartment was empty and so was my fridge. I had no car, so I decided to walk to the nearest grocery store, which was about a mile up the road. When I entered Harris Teeter, I was amazed at how large the store was. It was nothing like the small open-air markets I was used to in Baghdad.

I eventually found the cereal aisle and wandered down it. I looked left and right. Standing in front of the towering shelves of cereal boxes, I felt like a deer caught in headlights. A million different options, each one more colorful and tempting than the next. It was a stark contrast to the simplicity of my past, where choosing between Frosted Flakes or Mini-Wheats had been the extent of my breakfast dilemma.

I couldn't shake the feeling of being overwhelmed by choice, my mind reeling at the abundance before me. It was like that scene in *The Hurt Locker*, where the battle-hardened Staff Sergeant, accustomed to making split-second life-or-death decisions, is brought to a standstill by the seemingly mundane task of choosing cereal at the supermarket. That was me. Culture shock hit me like a tidal wave, leaving me paralyzed amidst the aisle of breakfast options. After what felt like an eternity, I finally snapped out of my stupor and reached for the familiar comfort of Frosted Flakes.

The next day, I returned to Harris Teeter to buy some beer. Surrounded by unfamiliar American brands, I grabbed a case of Heineken, carried it home, and nestled it in the fridge of my empty apartment. Despite the emptiness, I felt a strange fleeting sense of tranquility as I drank my cold beer, watching the rain hit the window pane outside.

It was a stark contrast to the bustling chaos of war, a peaceful interlude amidst the storm of life. As I stood there, watching the raindrops cascade down, I couldn't help but long for the presence of my family. I wanted them to experience this same sense of peace that was enveloping me.

To expand my perimeter, I'd go for walks or runs each day around the neighborhood. Qutt had introduced me to another Iraqi in the area, a devout Muslim who worked at Lowes hardware store. He'd help me with rides to the grocery store and would take me on errands. He didn't know I had long since abandoned Islam.

One day, when he stopped in, he found me with a beer in my hand.

"You don't need to drink that…" he said, startled.

"I'll stop drinking when I'm finished with it," I replied to him, the words tinged with a hint of defiance.

But then it struck me. This was the exchange I'd had with Bashar, and I couldn't help but share the memory with my new friend. At that moment, it felt as though I had come full circle.

18

NEVER STOP FIGHTING

I sat in my apartment, furnished now with a single bed, a chair, and small table, wondering what I was going to do next. I was grateful to be in America, grateful to have a second chance to create a life for myself. And yet I felt a deep sense of guilt from leaving my family behind.

When I left Iraq, I was under the impression that I would be able to bring them over to the States after I got citizenship. Yet I didn't know when that would be or what it would be like. During those first three months in the United States, I often wondered about ways to go back to Iraq to stay in the fight. I was in this weird limbo, trying to start a new life and yet feeling the guilt of leaving my brothers behind who were still fighting. Why had I made it out? Why was I given this golden ticket to freedom?

I was living in the land of the free and the home of the brave, but I found myself unable to fit in. My transition to the US was not going smoothly. While the stress of having a bounty on my head was gone, and the possibility of being shot was slim to none, something was still missing.

I struggled to interact with American civilians who seemed to constantly complain about 'first-world problems.' And I struggled to fit in with the Iraqi Muslim community since I had rejected Islam and religion altogether. I could not relate to either and felt totally alone.

I burned through my meager savings pretty quickly. America was much more expensive than Iraq. I tried to look for a job, but I didn't have any real work experience or connections so I was denied at most places.

I was down to my last few dollars, and soon I would not be able to pay for my apartment. The only skills I really had were my ability to speak English and my training as a soldier, which made the military my only viable path. With my thirtieth birthday looming, I knew I'd better join soon. I was getting old. I still wanted to kick in doors and liberate the oppressed. I really wanted to become a Green Beret. I missed my brothers. And I missed having a purpose.

I called an Army recruiter named SGT Camp whose number I'd gotten from Qutt, and told him I wanted to join in. I didn't have a car at the time but SGT Camp was kind enough to drive me back and forth to the recruiting station. During our car rides, we bonded over our time in Iraq.

He laughed, "Kind of odd for a new recruit to have more combat experience than his recruiter."

I got the job. It was really strange to join the Army at twenty-nine and already have six years straight of combat experience. I took my Armed Services Vocational Aptitude Battery (ASVAB) test to see what Military Occupational Specialty (MOS) I'd qualify for. But the problem was that I had learned the metric system that's used in basically every country but America. So when the test asked questions about yards and feet, I had no clue. Also, I'd never seen an American football field. In Iraq, a football field is a soccer field. Needless to say, questions with these cultural references put a dent in my performance.

However, I passed. My score was low, but ultimately it didn't matter because I was dead set on getting back into the action. My MOS was 11 Bravo, combat infantry. War was like a drug for this adrenaline junkie, and I was going to get my fix.

I had been given my orders to leave by August for bootcamp at Fort Benning in Georgia. But in the meantime, I had no way to pay for my apartment and was about to be homeless. Thankfully, Pete had introduced me to a kind family, Bill and his wife Debi. When they heard about my situation, they offered to have me move into their home for a few months; I didn't even have to ask. They showed up at my apartment and helped me pack the few things I had. They told me I could stay in their spare bedroom, if I wanted to. I'm forever grateful for their generosity and hospitality.

For those three months, I really became part of their family. I'd help out around the house and cook Iraqi meals for them. They loved my buttered chicken. I would even go out with Bill on handyman jobs to help him. He, of course, ended up teaching me so much about the trade, almost like an apprenticeship. It felt really good to be with a family.

One day, Bill went to his church to unclog the gutters. That was when I met Pastor Jim Sink. I'd never met a pastor before, so I didn't really want to engage in a theological conversation and prove him wrong at his church. I was an arrogant atheist, who typically loved to pick a theological fight to prove that people who believed in a 'Genie in Sky' were weak and delusional.

I thought, "He's a man of the cloth. I'll go easy on him this first meeting."

As I interacted with him that day, I was surprised by him. Pastor Sink was kind and dad-joke funny. His warmth and humor were a welcome change from the stern imams of my childhood.

On the way back, Bill and I talked a bit. I hadn't been able to help but overhear his conversation with Pastor Jim, where they kept saying, "God is this" and "God is that." In my head, I'd wondered, whether he really believed in all that? The more I had listened to their back and forth, the more curious I had become.

So I decided to find out.

"Do you really believe that religion?" I asked.

Bill looked at me. "I don't believe in religion," he replied. "I believe in a relationship."

I burst out laughing, and told him what I thought.

"All religions are the same," I said. "Man-made ways of manipulating people." I was adamant about proving my point.

Bill smiled. "Just talk to Pastor Jim," he said. "He would love to talk to you more."

I was emphatic. "No. I don't want to offend him or piss him off. He seems to be a good person."

Bill insisted. "Are you scared of the conversation or scared of the truth?"

I sighed. "Fine," I replied. "One conversation."

A few weeks later, I met Pastor Jim for lunch. He was trying to be culturally sensitive, so we met at a Mediterranean restaurant. But the food was terrible and not authentic at all. We started with a surface-level conversation when suddenly my anger towards God came raging full force. I didn't want to do pleasantries.

"If God really exists," I just blurted out, "then why does he allow all this evil out there? Why do good people get hurt? Why do innocent children die?"

I couldn't calm myself down, and yet Pastor Jim remained so steady as he sat across from me. And his answer caught me off guard.

"God never intended evil to happen to any of us, but Satan did," he said. "God's character and nature is love. God's original design was for everyone to be part of his family. That's why he created Adam and Eve in the garden. This place was full of beauty, intimacy, and freedom.

"But because God loves us, he didn't create us to be robots; he gave us free will to choose him or not. Unfortunately, when Adam and Eve listened to Satan and chose to eat off the tree of the Knowledge of Good and Evil, sin entered into the world. They chose to live without God. And we all came from Adam, and so we all have a sinful nature, a separation from God, that makes us want to go our own way. And a lot of people tend to do bad things and hurt others because of it.

"But that's why we have Jesus. God the Father sent his Son Jesus to die for our sins so that we could be restored to our original family, to a restored relationship with our Heavenly Father. Jesus forgives our sins, renews our minds, and changes our ways. He is a God of redemption."

I just listened. This was the first time I had heard the Gospel. The god of my youth, Allah, was not a loving father. There was no relationship; it was really just a religion with rules. This new concept was totally foreign to me. Going into the conversation, I had just wanted to win an argument, but clearly I hadn't. Somehow what Pastor Jim said made sense but I didn't want to admit it. As we left our meeting, I began to question everything. What if what he had said was true? What if I was wrong?

★ ★ ★

Each passing day brought me closer to shipping out for boot camp, and the thought of breaking the news to mom made my stomach turn. She had already been through so much, and I knew this would break her.

One evening, I finally mustered the courage to make the call. I sat on the edge of my bed, the phone in hand, my heart pounding. After a few rings, my mom's voice came through, warm and familiar, but with that hint of concern that always seemed to be there now.

"Hello?" she said.

"Hi, Yumma," I began, trying to keep my voice steady. "I need to tell you something."

"What is it?" Her voice tightened, sensing the seriousness in my tone.

"I enlisted in the Army," I said. The words came out faster than I had intended, as if rushing through them would make it easier. "But listen, this is a good thing. It will help me get my citizenship, and maybe it'll speed up the process to get you and my siblings' visas. Then we can be together again sooner."

There was a long silence on the other end. I could almost hear her processing the words, the reality sinking in. Then, suddenly, her voice broke, and she began to cry. The sound of her sobs cut through me like a knife.

"Why?" she asked through her tears. "Why did you leave Iraq just to go back to war? I don't care about the visa, Waleed. I care about you and your future."

I didn't have a good answer for her. I sat there, holding the phone, wondering the same thing. Why had I done it? For a better future, I had told myself. To fight for freedom. To be with my brothers-in-arms. For a chance to bring my family here. Honestly, I didn't fully know, and now, hearing her pain, my reasons felt insufficient.

"I love you," I said softly, my voice cracking. "I promise it will all work out, inshallah. If God wills it."

I hung up feeling like I had betrayed her, knowing I had added to her fears and pain. I tried to convince myself it was for the best, that this path would lead to something better for all of us. But deep down, doubt gnawed at me. Little did I know then that my citizenship and my veteran status would do nothing to expedite their visa process. As of this year, 2024, my family's visa cases are still pending. We have been waiting since 2011.

★ ★ ★

I arrived at Fort Benning on a steamy August afternoon. Very quickly, I realized I was the old man among a group of eighteen-year-olds, most of whom had never even held a weapon. Basic training was fun. And while 'fun' is a broad term in the military, overall I did enjoy the challenge and the camaraderie of it all.

After reception week, where we were issued our gear and processed into our training units, we began the first of three distinct phases of basic training. The first three weeks consisted of shock and awe. We were put under extreme physical and mental stress. The mental part was easy for me, but I struggled physically due to my more recently acquired beer gut.

Basic training was an interesting place to be both as an Iraqi and as someone who had already had a lot of combat experience with elite units. On the first day at sandhill, a training area with obstacle courses and fields, we went through our first smoke fest which was a series of brutal physical exercises. Then we went to our barracks. It was like something out of a movie. The room was full of bunk beds flanking the perimeter, and in the middle of the room there was a yellow line painted on the floor. This was where we had to 'toe the line,' stand in formation, receive orders, and get yelled at. Our only responses were to be "Yes, Drill Sergeant!" or "No, Drill Sergeant!" It was a place of discipline and strict order where we learned quickly that every action had a consequence.

One morning, our Drill Sergeants told us to toe the yellow line and sound off if any of us needed citizenship. Three of us responded.

"Reid from Jamaica, Drill Sergeant."

"Tala, from the Philippines, Drill Sergeant."

And then there was me.

"Hamza from Iraq, Drill Sergeant."

"WHAT!?" one of the Drill Sergeants shouted. "You're from Iraq? What the fuck are you doing here? What's your deal?"

"I was a translator in Iraq," I replied. "I moved here a few months ago, and I want to serve, Drill Sergeant."

After that interaction I was nervous how they'd treat me. The smoke fest continued but so did the conversations with my drill instructors. There was no special treatment whatsoever, but there was an understanding of my level of combat experience and perhaps a hint of suspicion or curiosity.

A few weeks into basic we had to do medical training. After each soldier finished their task, they would run up to the class instructor and shout out their rank, their last name, and the last four of their Social Security number so he could log their time.

It was my turn; I ran up to him and shouted my rank, my last name, and my last four.

"What?!" the instructor yelled. "What kind of a terrorist name is that?"

I instinctively went from parade rest to forming a fist.

"That's my fucking name, Sergeant!" I shouted, staring him straight in the eye.

But then I took a deep breath and tried to consider the context. This instructor was a medic and not a Drill Sergeant. And he had no patch under his flag, which meant he had never been deployed or seen combat.

"This asshole doesn't know shit," I told myself.

I relaxed my fist. I was determined to not let his racist remark get to me. In fact, I just wanted to prove all the more that I deserved to be here.

My goal was to get through basic training unnoticed. I didn't want the spotlight, and I didn't want to be discriminated against. I just wanted to fly under the radar and get the job done so I could move on towards my dream of one day becoming a Green Beret, like Pete and Dan and the other soldiers I had served with.

But the next day, the Company First Sergeant came to address the whole company.

"Where is Specialist Hamza?" he called out.

"Here, First Sergeant!" I shouted, as I ran toward him.

"Stand here!" he barked.

I stood in front of the entire company while he gave a speech.

"Specialist Hamza deserves to be here as much as any of you, if not more. He has already risked his life while in Iraq, serving our soldiers. Yet he has chosen to go and fight again, this time under the American flag."

He then concluded by naming me the company guide, which gave me the honor of carrying the flag from then on.

While anyone else would have been happy about receiving this honor, I was very nervous. I felt like I now had a target on my back. I knew the first Sergeant was using the situation with the medic to make a point. But I didn't want to be recognized because of someone's ignorance. I didn't like it, nor did I deserve it.

Sometimes I dealt with racism, but more often our run-ins were just comical misunderstandings. On one really hot day, we were outside eating MREs for lunch. I was obviously used to heat, and I knew to keep my hydration up by drinking water out of my camelbak as I ate. The Drill Sergeants always told us to drink water, but they also insisted on us keeping our camelbaks full.

I finished up my meal and sucked my camelbak dry. Just as I was picking up the MRE wrappers and stuffing them back into the bag to throw away, the Drill Sergeant barked at me.

"What do you think you're doing, Hamza? Why is your camelbak empty?"

Before I could respond, he marched me to my platoon Drill Sergeant, telling him my camelbak was empty. The Drill Sergeants were all sitting together in a group, among them, a new Drill Sergeant who did not know me.

He addressed me by rank.

"Specialist," he began. "Do you know what happens to you if you run out of water in Iraq?"

I smiled. All the other Drill Sergeants burst out laughing.

"Hamza, just go fill up your camelbak," one of them said.

I ran toward the water buffalo station to refill my camelbak and also started laughing. When I came back, the Drill Sergeant, who had tried to lecture me about Iraq and how hot it could be there, looked at me.

"Why didn't you tell me you're from Iraq?"

I smiled again and said, "I didn't want to embarrass you, Drill Sergeant." We laughed at the situation, and we moved on.

I loved range time but I got into a little trouble during basic training because I'd learned my shooting techniques from the SF guys. Their techniques were more advanced and prohibited at basic training. We were training on what's called bounding overwatch, using live ammo, which meant we would advance toward the target in pairs, covering each other in turn. One soldier would engage the target and the other would run to the next cover and engage the target to allow the other soldier time to advance, and so on.

Finally, it was our turn. The Drill Sergeant gave the command.

"Go!"

My buddy shot a couple rounds and then shouted. "Cover me while I move!"

"I got you covered!" I replied, and then I laid down suppressing fire.

Then my buddy moved behind cover and started shooting again, so I could advance.

As we continued this bounding pattern, my magazine ran out of ammo, and I needed to change it. I advanced again and signaled to my buddy that I was moving. As I sprinted toward my next cover, I flicked the empty magazine off my rifle, pulled out a fresh one, fed it into my rifle, released the bolt, got behind cover, and engaged the target. It was all muscle memory, honed from countless hours at the range at both Camp Justice and BIAP.

All of a sudden one of the Drill Sergeants started yelling at me.

"What the fuck are you doing?" he shouted. "Did we ever teach you how to do that? Why would you change your magazine like that? You were supposed to get behind cover!" And on he went.

He took me to my platoon Drill Sergeant and explained what had happened. He looked at me, visibly pissed.

"You better do what we are teaching you," he demanded. "Not that cowboy shit you did in Iraq. Do you understand?"

"Yes, Drill Sergeant."

"Drop, and give me twenty."

I dropped and did twenty push-ups. Fun times. I made sure I never did that on the range again.

The last three weeks of basic training were spent in Advanced Individual Training (AIT). We marched ten miles into the woods, spent a few days out there, and then marched back. To graduate, we had to complete all kinds of events during this last phase.

On our last night in the woods, I was on patrol around the camp. I moved quietly, careful not to make a sound. Suddenly, I stumbled into a hole, twisting my ankle. The pain was so intense that I fell to the ground, trying to suppress any noise. After a few minutes, the pain became unbearable, and I shouted for help.

Help came, and I had to be taken to the hospital. There, the nurse strung my leg up in the air, and I was told I had to stay in bed. Then the doctor came in to check on me.

"Son, you twisted your ankle so badly it's almost broken," he explained. "You really need to rest and rehab it."

"When am I supposed to do that?" I thought to myself, as disappointment washed over me. It was quickly followed by a tidal wave of frustration. And even though I'd still be graduating, completing basic training with an injury hit hard; the moment I had envisioned felt tainted.

Just before graduation, I received my orders to go to my duty station. I was to join the Third Infantry Division at Fort Stewart, Georgia. Or so I thought. Our cellphones had been confiscated when we first arrived at Fort Benning, and we only got a cellphone pass every two or three weeks. As graduation approached, I got a pass to call my friends and let them know about the ceremony. I didn't have any family in the US, but Dan's family had adopted me. His mother Georgia—who I called Mama G—and his brother came to the graduation, as did Pete.

After the graduation ceremony, Mama G and Dan's brother took me to IHOP. All I wanted was a good breakfast that didn't come from the chow hall. As we were visiting and eating, Pete called, asking where I was.

"I am coming to pick you up," he announced. "We are going somewhere."

"Okay," I said, as I hung up a little confused.

When Pete walked into the restaurant, I realized I had never seen him in his everyday uniform before. I knew he was a Green Beret, but in Iraq I never saw him wearing all his badges and tabs representing all the schools he completed. I was very impressed with all of his military accomplishments, which I now understood much better after going through basic training.

I thanked Dan's family for breakfast and for coming to graduation and jumped in the car with Pete.

"So, what do you want to achieve in the military?" he began as we pulled out.

"You know, Pete. I want to be the best. I want to be a Green Beret."

He then proceeded to tell me the level of commitment and focus that I needed to have to get there. Pete had become not only my friend but also my military mentor. I listened very closely to everything that he had to say, soaking it up like a sponge. After a few minutes, I realized we were driving back into Fort Benning.

"Ah...Pete?" I asked. "Where are we going?"

"The path I see ahead of you is a path that needs to be claimed. Otherwise you will not achieve your goal of becoming a Green Beret. You need to make things happen, and not just settle with the outcomes, if you want to achieve your desired results."

"Pete, where are we going?" I asked again.

"Ranger Battalion. We are going to the Ranger regiment."

"But Pete, I have a swollen ankle."

"You'll be fine," he smiled.

We arrived. Pete walked into Ranger Battalion and headed straight to headquarters. He talked to the operation commander while I waited outside. I started to see why he'd worn his uniform.

Then he finally came out of headquarters.

"I'm dropping you off at RASP," he explained. "The Ranger Assessment and Selection Program." Pete took his ranger tab off of his uniform and handed it to me.

"They will take care of you here. You are not going to Fort Stewart."

I looked down at the tab and made a promise to myself. I wouldn't let Pete down. I wanted him to pin that tab on me one day when I became a Ranger.

I would later learn that the Army rangers wore tan berets, and they have a special dislike for those wearing green berets. RASP was in an environment that I was not familiar with, but I quickly learned that here things were different from regular Army or even the Green Berets.

The first week in RASP hold, I realized I needed to get into better shape even though I weighed only 175 lbs and scored 290 out of 300 on the PT test. If I wanted to pass, I needed to be faster and stronger. But my ankle had started flaring up again, and when I tried to push through, I ended up twisting it once more.

My ankle set me back another month before I could finally start RASP. During that time, I dedicated myself fully to physical therapy—I wasn't going to let an injury stand in the way of my dream.

Finally, my ankle had healed, and I was ready to join the next cohort. The RASP hold building was across from the class building, and all new recruits had to get their gear and move to the next building. Sounds simple but the process was far from easy. We started at 2 a.m. and finished at 7 p.m.

There are two phases of RASP, each consisting of four weeks. After finishing the first phase, which focused on physical and mental conditioning, a group of soldiers including myself, were called in to meet with the class Non-Commissioned Officer In Charge (NCOIC). I knew that getting called in was not a good sign. I went through a mental checklist of all the assignments to see if I'd missed something.

The only thing I remembered was one time during push-ups when sweat dripped into my eyes, making it hard to see. I paused to wipe it away, and the Sergeant yelled at me and took my name. Aside from that, I completed all the events without issues and was never at the bottom of the pack.

I was really surprised and nervous to talk with the NCOIC. Each of us went in one by one. Finally, it was my turn.

"So what do you think of your performance?" he began.

"I tried my best, Sergeant. And I will continue to do my best."

He questioned me about certain events. Each time I would say that I did not achieve first in the class but I was not last. Then the questions changed, and the tone in the room shifted.

"Where is home for you now, and why are you here?"

It sort of felt like a trick question; I didn't have a home.

"I am here to be a ranger, and my home is here."

He continued down a strange line of questioning, insinuating that I might not be here for the right reasons and that my allegiance wasn't necessarily to the United States because I was Iraqi. So that was it. I was shocked, but I just kept answering the questions until he stopped abruptly.

The next words that came out his mouth were devastating.

"I am putting you on worldwide status. You can never be an active operator in the ranger regiment."

What?! I was so confused. To this day I still have no idea why I was kicked out of the Rangers. Was it because I was Iraqi, and there were still a lot of prejudices and distrust against us? The war in Iraq was still going on, and the US had lost a lot of soldiers. Or was it because I just wasn't a good enough soldier? I could handle not being a good enough soldier a lot more easily than my nationality not being good enough.

I called Pete the next day and told him what happened. He was pretty pissed.

"I'll come down and sort this shit out."

"Pete, I appreciate it, but I don't want to be a tan beret. I want to be in the Special Forces and earn my green beret."

"Are you sure, Waleed? I can help sort this out."

"Yes, I don't feel comfortable here. These are not my people."

"Okay. Well, there is a Special Forces recruiter at Fort Benning tomorrow. You can go talk to him."

The next day I went to see the recruiter. He pulled up my records on his computer and looked up at me from the screen.

"Why are you AWOL?"

"What, AWOL? I was in RASP, Sergeant." I replied. I knew that being Absent Without Official Leave was a big deal. He typed a few things into the computer and was quiet for a few minutes.

"It looks like your gaining unit at Fort Stewart never received you, and the Ranger Regiment never claimed you."

I was in shock. This was not how I wanted my military career to start.

"Give me a couple of days," he continued. "And I will fix this mess."

Sure enough, after a few days the recruiter changed my orders and requested a pending waiver. The waiver would permit me to go to the 'Q course' to train to be a

Green Beret. I needed this waiver because my ASVAB test score was lower than the required score.

After a few weeks, I still didn't have my waiver, so the recruiter sent me to airborne school at Fort Benning. He told me, if I got my airborne wings, I'd most likely be sent to Fort Bragg, which was the home of the Special Forces. From there I'd have a better chance at connecting with an SF recruiter. I had no other choice but to stay in the regular army.

After airborne school, I got my orders. I was indeed being sent to Fort Bragg in Fayetteville, North Carolina. My new unit was the 82nd Airborne Division. I was officially going to be a paratrooper. I had mixed feelings about that but at least Fayetteville was close to Raleigh, so I could go visit Pete, Bill, Debi and their kids on the weekend. I wouldn't be totally alone again.

I arrived at Fort Bragg on a rainy night, waiting for our unit assignments. We left our bags outside the building while we were being processed. After a long night of paperwork and waiting, I finally got to my temporary room only to discover that my backpack had been ripped at the top. Rain had been pouring directly into the laptop fan vent, causing my computer to die on me.

"Fuck!" I muttered in frustration. "Now I have to get a new computer."

With no money to spare, it was a terrible way to start my time at Fort Bragg.

During my first weekend off, I decided to head up to Raleigh to visit my American family and check out the bar scene. I hoped to drown my sorrows and maybe even pick up a good-looking girl. So I hopped on a Greyhound bus headed to downtown Raleigh, my sights set on City Limits Country Bar or the Hibernia Pub. My mission was simple:

drink enough to forget and find someone who could help me escape the pain, even if just for a moment. My previous marriage had left me bitter, with a deep distrust of women, and I had no desire to marry again. A one-night stand was the only commitment I was willing to make.

Eventually, this became my routine. For the first eight months that I was at Bragg, I'd head to Raleigh every Saturday to visit my American family. They'd invite me to church with them on Sunday, and I'd turn them down. As an angry atheist, I had no intention of ever stepping foot inside a church. Instead, I'd hit the bars Saturday night and then crash back at their house in the wee hours of the morning. I'd wake up around noon on Sunday to recover from my hangover before heading back to base. This was my norm, weekend after weekend, until one day, I finally gave in to their invitation to attend church—just so they'd stop asking me.

Walking into the church, I saw people singing and raising their hands. It was all very strange to me, but I noticed they seemed genuinely happy. I felt a strange sense of peace I'd never experienced before. I just sat back and watched. After the service, Pastor Jim approached me, officially welcoming me and saying he was happy I was there. I wasn't sure I was happy to be there, but I smiled politely and then asked if we could meet at some point to continue our debate from months ago. It was like inviting him to a duel.

He laughed. "Sure," he replied. "Next time you pick the restaurant."

★★★

I got my orders again. And this time, I was going to be deployed to Afghanistan to fight in Operation Enduring Freedom. Because I had been a translator and had an aptitude for languages, I was being sent to language school to learn Pashto, one of the languages spoken in Afghanistan.

Over the course of the next few months, as I prepared for my new mission, I began to have a different recurring dream beyond the normal; my weapon is jammed and I am stuck in a firefight nightmare. This dream was of three gremlin looking creatures in all black. They would look right at me and say, "Face us!" It was terrifying. I didn't have a clue what it was or why they kept visiting me in my dreams.

In hindsight, I can see that the more I rejected the idea of God, the more spiritual encounters I had. It felt as if God was relentlessly pursuing me even as I debated his existence with anyone who would listen. I wore my atheism like a badge of honor, a worldview forged in the chaos and carnage of war.

19

THE LADY IN RED

With just a few weeks left before my deployment, Bill and Debi invited me over to meet their friend Hannah. They said she was interested in hearing about my experiences in Iraq. Hannah was a filmmaker and a passionate researcher. Back in 2007, she had made a film called *Fire in Fallujah* about an American soldier who fell in love with an Iraqi woman during the war. Debi said she thought Hannah would find my story fascinating and that I could offer her some further insight into life in Iraq during the conflict.

  This particular Saturday I was lying on their couch, dressed in a t-shirt and my ranger panties—my really short army shorts—hungover from the night before. When Debi had mentioned her friend was coming for dinner, I just assumed it was a fifty-year-old woman and had no worries about my appearance. Then the door opened, and Hannah walked in. She was dressed in black jeans and a red shirt that contrasted strikingly with her fair skin. Her long brown hair flowed freely over her shoulders, framing her face with a natural elegance. I did a double take. My heart started pounding. I quickly popped up off the couch and excused myself. I ran up the stairs and confronted Debi.

"Why didn't you tell me your friend was young and hot?" I burst out.

"Why does it matter?" Debi replied, smiling. "I'm not setting you up."

Annoyed, I quickly changed into something more appropriate and ran back down the stairs. I smiled at Hannah and gave her a cheesy line.

"Excuse my twin brother," I said. "He is the naughty one."

She looked confused but smiled graciously. And that was when I fell in love. I know it sounds cliche, but time stopped for me. It was like I was looking at an angel.

We sat across from each other at the dinner table, the quiet clinking of silverware filled the air as we began to eat. I started telling Hannah my story, parts of it I hadn't shared with anyone else. As I spoke, I knew something had shifted inside me—my heart was now open. As she shared her own life experiences, I could sense the depth of her character—her kindness, her strength, her wisdom. This wasn't just physical attraction anymore. I knew she was someone of true value—someone I could never find in the shallow bar scene. This was something far more meaningful. This was different. She was different. It was love.

Unfortunately for me, the feelings were not mutual. Hannah was completely focused on hearing my story at dinner that night. She wanted to know about the war, my trauma, and the historical and religious complexities of

Iraq's cultural divide between Sunni and Shia Islam. She was clueless that I already had feelings for her. I knew I was about to be deployed to Afghanistan, with a real possibility of dying in combat, so I decided that—if I made it back and she was still single—I would pursue her.

On March 20, 2012, around 10:30 p.m., my unit, the 82nd Airborne Division, and a portion of Bravo Company 1-504 PIR, loaded into big Bluebird buses and headed to the tarmac at Pope Airfield. There, we loaded into private jets that the DOD had chartered for us. It was a unique privilege to get on a civilian airplane with a M4 carbine rifle in my hand.

As our gear was being loaded under the aircraft, I could sense the excitement of the other soldiers. Meanwhile, I was having a very different reaction. Nervousness coursed through me as I carefully folded a letter I had just written to Hannah. It was for her to read in case I didn't make it back. I wanted her to know how I felt about her, to understand the depth of my feelings, if the worst were to happen.

As I looked out of the window, all my memories of Iraq started flooding back. The plane began to take off, and I closed my eyes and wondered what the fight would be like in Afghanistan. My war experience in Iraq had involved running around with the cool SF guys fighting in urban areas. In Afghanistan the ROE's would be different; the terrain, the people, everything would be different.

I looked down and pulled my dog tags out of my shirt, pondering the inscription on them. *Religion: Muslim.* I couldn't shake the idea; I really had no religion. If I died in combat, I had requested only two things: to be laid to rest in Arlington National Cemetery with my family in attendance and the song "Fade to Black" playing as my casket is lowered into the grave.

We arrived at Manas Air Base in Kyrgyzstan sometime in the evening the following day. The squad leader called us together to get a headcount as we waited for orders and temporary housing. I looked to my left and spotted a group of SF guys. You can always recognize them when you see them, because they're not subject to the usual rules for military attire. One or two might be wearing baseball caps or even have a beard.

I wondered if I knew any of them from Iraq. I looked closer.

"Wait, that's Matt!" I blurted out, and I quickly walked toward him.

I could see my squad leader out of the corner of my eye, monitoring my every move. But I was laser-focused on Matt.

"Matt!" I called out. "Is that you?"

"Waleed!" Matt replied, grinning. "Holy shit dude, what are you doing here?"

"I'm here with the 82nd," I explained. "We're going to Ghazni."

"Dude, I can't believe we are both going into Afghanistan at the same time. Come meet the guys."

Matt introduced me to the rest of the team and told his team CPT how we had met in Iraq, and I had been one of his terps.

"Hey, dude, do you want to come work with us?" Matt asked.

"I wish I could, but I don't know if I can do that," I answered. "My orders are to deploy with my unit."

"We can pull you in," Matt said confidently. "You're 09 Lima, right?"

"No," I replied. "11 Bravo."

"We can't pull him, because he is not a terp," the team CPT chimed in. "He is a US soldier."

Shit. For a few seconds I thought I was going to relive my glory days with the SF guys. I headed back towards my unit, thinking that, although I might not be allowed to fight with the old team, I was still proud to be a paratrooper. And I was just going to hope Afghanistan wasn't that bad.

After a couple of days in temporary housing, we loaded up into a C-17 GlobeMaster aircraft headed to Bagram Airfield. It was standard protocol for anyone coming into the country to spend a few days doing mandatory training and zeroing of weapons, calibrating them for accurate aim. During this training stint, we lived in a tent that was big enough for about 200 to 300 soldiers. It was almost full and an impressive sight.

On March 31, eleven days after we had first loaded into buses in North Carolina, we left Bagram and arrived at FOB Sharana. A few hours later, we loaded up again, this time in Chinooks headed to FOB Warrior, in Ghazni, near the Pakistani border. Our journey was almost over. Two days later, we got into MRAPs and drove south towards Muqur where our COP was located.

As we neared the city, I noticed a green road sign.
Welcome to Muqur.

"Home, sweet home," I thought. After fourteen days of traveling, it all felt like something out of the movie *Trains, Planes and Automobiles*.

Our beautiful home for the next six months consisted of a few AirBeam tents and a couple structures near the Tactical Operation Center (TOC), all surrounded by blast walls made of double-stacked Hesco bags. When we first arrived, the chow hall wasn't operational yet, so we were living off of MREs. These 'Meals Ready to Eat' have enough nutrition to keep a soldier going, but they're no one's idea of delicious cuisine. A few weeks into our stay, the kitchen was finally up and running, but the food didn't get much better. We ate turkey patties and mashed potatoes with cumin and hot sauce for breakfast. Then we were forced to skip lunch, and dinner would then be the same as breakfast. I was starving. Finally, a month into our deployment, the finance guy showed up, and we were able to withdraw cash. Now we could finally buy food and cigarettes from the locals.

Our translators were our lifeline to the outside world and essential in securing local food and supplies. Having been in their shoes back in Iraq, I understood their challenges and approached them with compassion and a strategy. I'd hand one of them cash, asking him to buy food. But I'd always tell him to get enough for himself as well. And I always made sure there was a little extra left over as a tip.

There were a couple reasons for this. The first was that I knew what kind of risk he was taking just by being our interpreter. And secondly, I wanted to make sure that the food was not poisoned. It was always a good idea to eat what your terp ate. And thankfully, the local food was delicious. The bread—the naan—reminded me of my grandmother and the times I used to steal it fresh out of the oven.

As soon as we entered COP Muqur, we hit the ground running, establishing our presence and making connections with the locals. Entering this remote part of Afghanistan felt like going back in time a hundred years.

My first mission was to patrol a village called Mu'men Kheyl, full of farmland and mud huts. We were outside the wire patrolling unfamiliar territory amongst unfamiliar people. I now knew what the US soldiers must have felt like in Baghdad. For me, it had been home; for them, they hadn't had a clue. There was a certain uneasiness that always came with patrolling an area you didn't know. Now it was my turn to feel it.

Memories of missions in Iraq came flooding back to me during this first patrol. I was filled with a fond sense of longing for my friends. But I had to remind myself, most of them were dead or had moved on to their next mission. I shook my head, trying to snap out of it.

"I need to keep my eyes peeled and my head on a swivel," I thought. "Or else I am going to get myself or someone else killed."

We were pushing deeper into the farmlands when suddenly the sharp crack of AK-47 fire echoed through the air. We quickly jumped into a ravine a few feet away and started to return fire. I grabbed the radio and reported back to the TOC to give them a Situation Report (SITREP). We quickly gained fire superiority, and the Taliban retreated. I thought they were mainly focused on testing our response in order to hit us harder the next time. But, nevertheless, our Platoon Sergeant boasted about our success.

"Looks like all you motherfuckers will be getting your CIBs," he told us.

The Combat Infantry Badge was given to soldiers who had engaged in a firefight with the enemy. That much we definitely had done.

Our mission at that time was pretty clear: identify areas of interest and possible Taliban targets. We worked alongside the Afghan National Army (ANA) and continued to patrol the villages in our AO. After a few patrols as Radio Tactical Operator, I was assigned to help our platoon leadership because of my experience as a terp and my ability to speak Pashto. Depending on our target's nationality, I would translate either from Pashto or from Arabic for our lieutenant.

A few months later, we were part of a large operation to gain control over several villages east of our COP, across the Kandahar-Ghazni Highway. The operation involved two companies, and our mission was to infiltrate the villages from the north and sweep through them to the south.

At 2 a.m. we left COP Muqur on foot, exiting through the back gate and marching north. I was the third man from the front, part of the first squad Alpha team. As we approached the highway, I could see our crossing point: a small tunnel that passed under the two-lane road. Our spread-out formation squeezed through the tunnel for concealment then continued marching another three to four miles until we reached our starting position.

It was still dark when we arrived at our staging area just a few hundred yards outside the village. And slowly, daylight began to break as we quietly progressed through the village.

Bang! Bang! Small arms fire started.

We engaged and began to fire back. We had encountered this mission's first firefight.

The next three to four days, we engaged in multiple firefights as we continued through the villages. Once we arrived at the last village, a few MRAPS were waiting to pick us up and bring us back to the COP.

We went straight to our tent when we arrived. Our platoon leader joined us.

"It's time for our AAR," he began. "And y'all need to kiss your asses. EOD just cleared a 400-pound IED buried in that tunnel you went through when you passed under the highway. If it had gone off, you all would have evaporated."

We were all in shock. How did we miss a 400-pound IED? My mind was reeling with questions.

"Why didn't it go off? Why didn't I die?" I couldn't help but wonder.

This was another divine intervention.

A few weeks later, I called Bill back in Raleigh, and I told him about the IED.

"Pray to whomever you pray to," I said. "Because I don't think I'm coming back. The fighting here is different from Iraq. Iraq seemed easier. Maybe it was because I was familiar with the urban environment. I don't know, Bill, but I really think I might die here."

"Don't worry," he responded. "You will be okay. We are all praying for you here. God's got you."

I rolled my eyes, and then I asked him a favor.

"I'm starving. Can you please send me a care package with some snacks like beef jerky? And I really need some tobacco pouches."

As Bill hung up, he promised to send a care package.

It was always like Christmas morning whenever a care package arrived. I was so grateful to Bill, to the USO, and even to other soldiers' moms who adopted me and would send packages from time to time. Every package was a dose of encouragement to keep fighting and come back alive.

Finally, my six-month tour was over, and I was ready to leave Afghanistan. I couldn't wait to get home and see if Hannah was still single. I was aching to talk with her. I missed her so much that I had named my rifle after her and had slept with it every night.

I also looked forward to receiving my Combat Infantry Badge (CIB), as well as my unit patch, which would both permanently connect me to the history of the 82nd

Airborne Division. These symbols of my deployment would finally make me look like a real soldier, not just someone fresh out of basic training.

We were just waiting to get our orders to leave when we started hearing through the Private News Network (PNN)—also known as the rumor mill—that we would be flying out of FOB Warrior. I thought this would have been an easy exit, but that ended up not being true. You can never trust the PNN.

Unfortunately, our exit was not so simple. Instead, we were instructed to get all of our gear and load into the MRAPs. We then drove overnight to the base in Bagram, passing through several areas known to be IED hotspots along the route. We had to dismount periodically and check for them on both sides of the highway, using handheld mine detection equipment. All I could think about was how badly it would suck to make it through the entire deployment just to get blown up on my way out.

I felt an overwhelming sense of relief when we finally arrived in Bagram. It was a step closer to civilization, a welcome change from the constant tension of the warzone. Bagram had everything I hadn't realized I'd been missing: fast food restaurants, an incredible chow hall, and even catalogs from which you could buy a Harley Davidson motorcycle or a brand-new Dodge Dart and have it shipped right to your door back home.

It was a brilliant sales strategy. We had all saved up our hazard pay and felt a sense of accomplishment just for making it through the deployment. With money to burn and a seize-the-day mentality, it was hard not to get caught up in the excitement. I felt like a kid in a candy store, tempted by the idea of cruising back home on a shiny new Harley.

But then, I remembered that I had made a promise to my mother. The only thing that held me back from splurging on the motorcycle was the thought of seeing her again. I had promised her that, if I made it back safely from deployment, I would fly her to Istanbul. We'd spend ten days together, exploring the city, seeing the sights, and catching up after three long years of being apart. That promise kept me grounded, reminding me of what really mattered. In just a few weeks, I'd finally see her again.

20

THE MAN IN WHITE

As soon as I landed back in the United States, I could not wait to call Hannah. I wondered if she had found a boyfriend over the last six months. I didn't think so because I had been Facebook-stalking her all during deployment whenever I got my turn to get online. Unless of course, she was dating someone and hadn't made it Facebook official?

But I figured, "What do I have to lose? I've almost died a million times. At this point, I can handle a girl rejecting me. It's worth a shot."

In the barracks, I sat on my twin bed with a blue and white striped comforter, the kind that starts rough and stays rough no matter how often it's washed. I'd spent so many nights just laying on this blanket and not sleeping, because the moment I'd close my eyes the nightmares would inevitably begin.

I dialed Hannah's number.

"Will she pick up?" I wondered. "What will I do if I get her voicemail? Shit." I hadn't thought this through…

Whew. She answered. I didn't have to navigate the dreaded voicemail and leave some awkward message. I was relieved. And jumped right in.

"Hey, Hannah, uh… I just got back from deployment, and I'd love to tell you all about it. Are you free Friday night for dinner?"

I knew she was a storyteller so that might entice her to say yes. My strategy was: no pressure—come just listen to my story again.

To my amazement, she agreed.

When Friday night came, I met Hannah just around the corner from her office, at the Rockford, a charming neighborhood spot. It was the kind of place where people had been gathering for years, enjoying hearty food and warm conversations, making it a beloved local institution. It was a warm summer night, so we sat out on the tiny rooftop patio. I was dressed in a button-down shirt and nice khaki pants—very different from when I had met her the first time. The conversation was pleasant, but midway through, I leaned in very intently.

"So, I like you, and I really want to date you," I declared.

She looked at me wide-eyed, her face starting to blush.

"Shit," I thought. "I hope I wasn't too forward. I just don't want to waste time playing games."

She grabbed her drink and took a sip to clear her throat.

"Please excuse me. I need to run to the restroom."

She slipped away, and I sat there, trying to remain composed, but my mind was racing. Had I scared her away? Did I give off stalker vibes? When she returned, she looked nervous. She took a deep breath and then leaned forward.

"I know you probably won't understand this, but you and I will never work out because my faith in Jesus is at the very core of who I am. And every decision I make flows from it. We'd never be compatible since you and I don't have the same faith."

She was trying to gently put me in the friend zone while staying true to her convictions.

I sort of laughed. "Well, I figured," I said. "But I had to try."

As we parted ways, she pulled a Chandler Bing from *Friends*, saying something like, "We'll do it again some time, friend!" with her nervous laugh.

After Hannah left, I found myself wandering down Glenwood Avenue, lost in thought. Who am I going to end up with? I really did love Hannah, but I couldn't tell her that; she'd think I was a stalker. It was an odd thing to explain to someone without freaking them out—that you loved them when you didn't even really know them. But I wasn't going to give up that easily. I resolved to keep pursuing her, convinced she would eventually come around. She was worth fighting for.

★ ★ ★

A few days later, I collapsed onto my bed, still dressed in my uniform, exhausted from a long day of training. I closed my eyes and drifted to sleep. In my dream I saw a man standing before me, clothed in white. His face radiated such brilliance that I couldn't discern any features. He looked at me and spoke these words.

"I am. I am. I am."

I was completely freaked out. Since leaving Iraq, I had suffered with severe PTSD and had nightmares every night. My nightmares always had the same theme—I am in an ambush or in an intense firefight, and I am out of ammo. Or my rifle is jammed or I've lost my leg. And I can't shoot back. I'm stuck. I see my deceased friends in danger, and I can't protect or save them. Then I hear the mortar impact explosion, and I wake up, usually in a pool of sweat with my heart racing. My typical night consisted of trying to stay up as late as possible because I knew what would happen the moment my eyes would close. Every night I was fighting the enemy of sleep.

But I couldn't shake this dream. Who was this man in white? It wasn't like my other dreams. I decided I was going to figure out who he was and what he was talking about. At this point I had no reference for Jesus or the Bible, so I was genuinely curious what this dream meant.

Shaken up, I walked out to the breezeway in the dim light of evening to smoke a cigarette and calm my nerves. My friend Marrow joined me. We had become close during our language school before heading to Afghanistan. He could tell something wasn't right and asked me if I was okay.

I thought maybe he would have some insight into this dream. I shared the dream with him. His face erupted into a smile, and he gave me a hug.

"Jesus loves you, bro." he said, genuinely.

I pushed him off, confused and irritated.

"Fuck off!" I retaliated.

Pissed, I walked down the stairs heading towards the open area by the barracks.

I looked up at the night sky. The stars were twinkling, my mind still racing. I kept wondering about the man in white. Suddenly, I felt a surge of fear and just wanted to forget about the dream. I sat down on the ground, leaning back against the brick wall of the barracks, and lit up another cigarette. As the nicotine hit my veins, I started to calm down, and my mind wandered to Hannah.

The friend zone is never a good place to be, and I was determined to make a comeback. Since Hannah hadn't given me a flat-out 'no' that meant I still had a chance. It occurred to me that we had an Army ball coming up, a post-deployment celebration, and I thought a ball might be just the thing to get me out of the friend zone. All I needed was to convince Hannah to go with me.

Easier said than done. The next day I called her, and she politely declined.

"Thank you, but we are just friends," she replied. "And I don't want to lead you on."

"Shit! The dreaded 'friend' word again." I thought.

Looking back, I can see Hannah was genuinely trying to guard her heart and didn't want to hurt me. She just kept saying the ball felt too high pressure. Maybe she worried it would turn into one of those fairy tales where we'd fall madly in love as we danced the night away. Little did she know, that's exactly what I was hoping for.

But there were things happening behind the scenes, and a week later Hannah called me back. Ever since Hannah's mom had passed away from cancer a few years ago, her older sister Jenny had become her go-to confidante. Hannah later told me I had Jenny to thank for her change of heart; apparently, Jenny talked Hannah into finally going to the ball with me. Jenny had given her the ultimate guilt trip.

"Waleed served our country and almost died." she argued. "The least you can do is go to the ball as his friend." Then she ended it with a mic drop. "Have you at least prayed about it?"

When I saw Hannah's number on my phone, I picked up, forcing myself to be extra casual. Reluctantly she told me she'd changed her mind. She would go to the ball with me. I remained calm, but on the inside I was freaking out. What was happening?

During the entire next week, we went back and forth, arguing about who was going to drive.

Hannah's voice was always soft but her tone remained resolute.

"I really think it's best if I drive myself, Waleed," she said a few days before the ball. "Fort Bragg is an hour and a half away from Raleigh, and I'd like to be able to leave whenever I want to. You know, just in case."

As a military man, I knew exactly why Hannah wanted to drive. She needed an exit strategy. But I also wanted the extra three hours in the car with her. So I took a deep breath, trying to keep my frustration in check.

"Hannah," I began. "Fayetteville isn't the safest place to be driving alone at night. I don't like the idea of you being on the road late by yourself."

She sighed, the sound of her exasperation clear even over the phone.

"I appreciate the concern, but if I can travel the world by myself, I think I can handle an hour and half drive. I'll be fine."

"Please, Hannah," I pressed, trying to inject sincerity into my voice. There was a pause, and I could almost hear her weighing her options on the other end of the line. Finally, she spoke, her voice softer.

"You're really not going to let this go, are you?"

"Nope," I replied, "I'm stubborn like that."

She let out a reluctant laugh.

"Alright, fine. You win. You can drive. But if I want to leave early, you have to promise to take me home without any arguments."

"Deal," I agreed quickly, relief flooding through me. "Thank you, Hannah. This really means a lot to me."

"Yeah, yeah," she said, her tone teasing now. "Just don't make me regret this, okay?"

"I won't," I promised, my heart racing with excitement as I hung up the phone. This was happening! I was one step closer to moving out of the friend zone.

The night of the ball, I drove to pick up Hannah in my green 2003 Hyundai Santa Fe, which I had proudly named LaShonda. I loved this car, especially because I had been able to buy it with my deployment money, even after getting the tickets to see my mom in Istanbul. I had made sure it was spotless inside and smelled nice. I even had a Christmas tree air freshener hanging from the rearview mirror.

Decked out in my army dress blues uniform, I stepped out of the car, grinning from ear to ear. My heart raced with anticipation as I knocked on Hannah's door. When she opened it, my jaw dropped literally. She was wearing a stunning, long, black dress, her dark hair swept up elegantly.

She blushed at my reaction, quickly tapping my jaw shut.

"Okay, you ready to go?" she asked, a hint of embarrassment in her voice.

I snapped out of my daze and managed a smile.

"Yeah, let's go," I said, offering her my arm as we walked towards LaShonda.

When we got into the car, I handed her a small jewelry box. Before she could decline the gift and remind me this was not a date, I interjected.

"This is just a thank you. It's what we do in my culture."

She smiled and opened the box; it was a beautiful dragonfly necklace. She looked surprised.

"Did you know dragonflies were my mother's favorite?"

"I had no idea." I replied, sincerely. But I hoped this would get me some brownie points.

I started driving towards Fayetteville, the hum of LaShonda's engine filling the silence. Trying to avoid the awkwardness of this non-date, I broke the ice with some small talk.

"So, what kind of music do you like?" I asked, glancing over at Hannah, who was looking out the window.

She smiled, her eyes lighting up. "Oh, I don't know, I like a lot of music. Singer-songwriter stuff, but I'm also a sucker for anything '90s. How about you?"

"Metallic. Metal and classic rock." I quickly blurted out.

We traded favorites back and forth, laughing about guilty pleasure songs and sharing memories attached to certain tracks. The conversation flowed easily, but there was still an undercurrent of tension, a sense that we were skirting around something deeper.

Then, as we passed a sign for Fayetteville, Hannah shifted in her seat and looked at me with a seriousness that caught me off guard.

"Waleed, can I ask you something?"

"Of course," I replied, trying to keep my tone casual, though my heart rate picked up a notch.

She hesitated for a moment, then plunged ahead.

"Do you have PTSD?" she asked. "Do you have nightmares or dreams?"

Her question hung in the air, heavy with unspoken curiosity and concern.

I hesitated for a moment. I was trying to impress this girl and had no intention of opening up about my struggles with mental health or PTSD. I took a deep breath, my grip tightening on the steering wheel. I decided it was best to just tell her about the dream I had recently.

"Well, just the other day I had a strange dream. There was this man dressed in white, saying 'I am, I am. I am.' Three times. It was so strange. He was so bright, I couldn't even see his face."

She looked at me and smiled.

"What?" I asked.

At this point, she knew I had been a Muslim but currently considered myself an atheist and had no knowledge of the Bible.

"Waleed," she said gently. "That was Jesus. In the Bible it says he is the 'I am.'"

As she said the name of Jesus, I felt rage start to rise inside me. I know now I was starting to manifest demons but at the time all I felt was turmoil being stirred up. I looked at her, filled with anger.

"Fuck that!" I spat vehemently, along with a few other choice words. I was coming unhinged.

Hannah just got quiet, her eyes glued to the passing scenery outside of her window. This was not how I had planned the evening to go, but I couldn't snap out of it. As we drove, the tension mounted. I could tell Hannah was regretting not driving her own car and was trying to figure out an exit strategy.

Finally, after several minutes, I calmed down.

"Hannah, I'm really sorry for yelling at you and turning into a raving lunatic. It's just…"

"What, the name of Jesus?"

Hannah knew that she had struck a nerve and that the demonic was rearing its head. But she also knew the authority that she had as a believer. And later she told me that she had been praying in the spirit, casting out every demon, during that time of awkward silence.

"All of it," I replied. "I don't know what is happening to me. I have a lot of questions, and growing up Muslim, I don't have a reference for this. I'm just really confused."

"That's okay, and I don't have all the answers either," she said graciously. "I didn't grow up like you, but what I do know is that God is pursuing you, Waleed. He always has been from the beginning of time. His hand was on you all those times you should have died but didn't. You don't know this, but when you were in Afghanistan, I prayed for you every day. I prayed that God would protect you and that you would come to know Him."

The tension finally eased as we arrived at the ball. But I couldn't shake what Hannah had said. God was pursuing me. If this was true, was that dream God's way of revealing himself to me?

We had the best time at the ball. I don't remember much about the food or the event, just that we spent most of the time on the dance floor. While we were dancing to "Livin on a Prayer," Hannah leaned towards me.

"I had no idea you could dance," she whispered. "I love to dance, but most guys just stand in the corner nursing a beer all night. Or, at best, do the two-step sway."

I laughed as I went down low to do the Russian kick, almost splitting my pants. We both started cracking up.

The ball was ending but I didn't want the night to be over.

"Do you want to keep dancing?" I asked, hoping I wasn't pushing my luck. "I know a place we could go."

"Sure," she replied. "What do you have in mind?"

"Don't worry about it. It's a surprise," I said with a playful grin, eager to see her reaction as I pulled up to a country line-dancing bar.

"Have you ever been to a cowboy bar and line danced?"

"No," she laughed. "I'm a city girl. Have you?" she snapped back.

I just smiled and grabbed her hand.

"Come on," I replied, as I led her into the Cadillac Ranch.

It was a dive bar with corrugated metal walls and exposed wooden beams in the rafters, exuding a rustic charm. The dance floor, the heart of this honky-tonk, was larger than life, buzzing with energy and the sound of country music. Surrounding the dance floor were high-top wooden tables, perfect for setting down your beer while you took a break from dancing.

It was a funny scene: an Iraqi army guy in a country cowboy bar with a city girl. Neither of us looked like we belonged. But that's where she was mistaken.

The song "Copperhead Road" came on, and I grabbed her hand, twirling her onto the dance floor. I started stomping my feet, doing all the moves to the song while she stood there, dumbfounded. Who was this Iraqi-country-army man? She must have thought. It was exhilarating. We danced until the sun came up, and it wasn't until then that she insisted it was time for me to drive her home.

My plan had been to sweep her off her feet with my dance moves, confident that I'd impress her enough that night to make some significant strides out of the friend zone. But to my surprise, Hannah stood firm in her convictions, reminding me all the way home that we were still just friends. But I wasn't going to give up that easily.

A week later, Hannah called me and asked me to pick her up from the airport after a work trip to Europe. She told me she had something to give me. I agreed, inwardly rejoicing.

"Hell yeah!" I thought. "I'm finally out of the friend zone."

With anticipation, I arrived at the airport early, checking the boards inside the arrival terminal every ten minutes to see when her flight from Paris to RDU would land. I wondered if I should grab some flowers. Or would that give off creeper vibes? I opted against it and thought it's better to play this low-key so I didn't scare her off.

As soon as Hannah's plane landed I sprinted to the baggage claim area where I had promised to meet her. After a few minutes, she rounded the corner, smiling. I nonchalantly walked up and gave her a hug. Then, we waited together for a few minutes for her bags to arrive on the carousel.

"Want to grab some dinner?" I asked casually. "I know you've been flying for twelve hours."

"Sure. I'm starving."

I quickly grabbed her bags off the carousel, and we walked towards LaShonda. I drove to a local cafe that I knew had gluten-free options in the hope of impressing her. If it were up to me, I would have just gone to Five Guys and gotten a burger and fries. However, I knew this was going to be a special evening, and I wanted her to know I was attentive to her dietary restrictions that she had recently shared with me.

We sat down and ordered a few dishes. She shared a little bit about Paris, seeing the Eiffel Tower and eating the most amazing GF crepe. Then she pulled out a book, *Jumping Through Fires* by David Nasser, and slid it across the table.

"I don't know if you are much of a reader, but I figured you'd be stuck on a plane for fourteen hours on your way to Istanbul, so you might want a good book."

I picked it up and glanced at the cover.

"Oh, but don't read Chapter 13," she continued seriously. "It's about when he meets his Southern Christian wife. I don't want you to get any ideas or to lead you on. I'm giving you this book because I really feel like David—who was an Iranian Muslim—had so many of the same questions you do. And I think it could help you on your journey to find the truth."

She had done it again. The words about 'not leading me on' were like a knife in my chest that she was slowly twisting. It took me a second to recover and then I smiled, thanking her for the book. I don't remember the rest of the evening except that curiosity was getting the better of me. Her well-meaning comment had been the spark. What was in Chapter 13? It was like a forbidden fruit, and I couldn't help myself.

Ten hours later, I was on a plane bound for Turkey. The long-awaited reunion with my mother was finally within reach. Yet, curiosity took over. As soon as I settled into my seat, I pulled out the book and flipped straight to Chapter 13. What I read there intrigued me. David's wife sounded just like Hannah, so I decided to flip to the beginning and read the book—to figure out how he managed to marry her. I read the entire book on the plane and then twice more while I was in Istanbul. It was captivating at a level I couldn't quite explain.

The moment I saw my mother at the airport, a wave of relief and joy came over me. Three long years and a war had stood between us, but now, none of that mattered. We rushed into each other's arms, and she broke down, sobbing against my chest. I held her tightly, feeling the weight of every lost moment melt away.

After what felt like an eternity, she pulled back and wiped her eyes, managing a slight smile.

"Waleed, you've gotten so fit," she said, her voice a mix of pride and disbelief.

I laughed. "It's all those army workouts."

Standing there, reunited with my mother, in a place where we didn't have to look over our shoulders, didn't have to fear for our lives—it was a feeling I couldn't put into words. For the first time in years, we could just breathe.

After our day touring the sites and enjoying the cafés, we spent the night at a hotel near the old-city. I was jet-lagged and couldn't sleep, so I stayed up rereading the book. Initially, I was drawn to David's story because I was searching for the formula to get out of the friend zone. But then something shifted. While rereading the pages night after night, I began to identify with his struggles. And his struggle to surrender. Reflecting on my own journey, I sensed I was at a crossroads. For the first time I could see clearly; the fog of war was lifting. God was peeling back the layers of my life, one at a time.

First had been my disillusionment with religion during the war. I questioned Islam and how a good god could allow so much pain and carnage. Especially how he could let people commit so many acts of violence in his name.

Then I had been removed from my family and culture. When I arrived in America, I was an angry atheist. Then came my near-death experience in Afghanistan and then finally the dream of the man in white. I had started to ask questions.

"Why was I still alive? Who was this man in white? Was it really Jesus? Could I ever feel lasting peace like I'd read about in this book?"

These questions had begun to grip me and weren't letting loose.

A battle raged within me as I lay on the hotel bed, book in my hands, reading by the glow of my phone's flashlight. My entire worldview was unraveling as I consumed the pages of *Jumping Through Fires*.

I put the book down and glanced at the clock. It was 3 a.m., and my mother was sound asleep beside me. I quietly slipped out of the room, careful not to make any noise, and headed to the hotel lobby. My chest felt tight; it was like I couldn't catch my breath. My heart was pounding, and my mind was swirling. All the questions wouldn't stop.

I needed to talk to someone, anyone, to make sense of what I was feeling. I found a seat in a quiet corner of the lobby and pulled out my iPad. I opened FaceTime and called Hannah. It was still early in the States, so I hoped she'd pick up. I needed her perspective. After all, it was her fault I was feeling like this—she was the one who'd given me the book!

Thankfully, she answered. I tried to play it cool at first, telling her about the grand bazaar and the ancient sites we'd seen the day before. But then she asked me about the book. I was just about to open up to her about the struggle I was having when my mother sat down next to me. Why she's woken and how she'd known to look for me in the lobby, I'll never know.

But there she was. She smiled and asked me in Arabic who I was talking to so late. I told her it was just a friend. My mother snatched my iPad enthusiastically and looked right at Hannah. She waved at her and then spoke in Arabic.

"Hi-bitch," she said.

Hannah looked mortified, thinking my mother was saying a curse word. But then Hannah smiled."Hi-B…" she started to reply.

But I interjected. "That means I love you in Arabic!"

To her credit, she simply nodded and started again.

"Hi-bitch," she replied in her best Arabic accent.

Crisis averted.

The next day, my mother and I were touring the famous Hagia Sophia, a mosque that had once been a Christian church. As I walked through its grand halls, I couldn't help but notice the remnants of ornate gold crosses painted on the ceiling. Now they were covered with the newer black paint of the crescent moon of Islam. The sight struck me. This strange dichotomy resonated with me, the two ancient religions coexisting in the same space, each layer telling a different story.

As I looked up at those two vastly different symbols —the cross and the crescent—a wave of familiarity washed over me. My whole life had been a dichotomy. I had for so long been caught between two worlds, much like this building. One symbol, the cross, represented surrender—a faith rooted in sacrifice and grace; the other, the crescent, represented conquest—a belief grounded in rules and power. Deep down, I knew what was true. But standing there, in that place where history and belief collided, I wrestled with accepting it.

Enamored by the ancient city, my mother wanted to go pray inside the historic Blue Mosque, which was just around the corner. I followed her towards the mosque but stopped just shy of the entrance. I stood under the large archway which framed the courtyard just before the entrance.

Then I tried to take a step forward into the mosque, but I couldn't move. Was someone pulling on my shirt? I looked back but didn't see anyone behind me. I tried to take another step, but I still couldn't move. Unsure of what to do, I told my mom to go ahead inside without me.

As she walked into the mosque, I turned around and looked up to the sky.

"If you are real," I whispered, "I will believe in you. But you have to do the rest."

I had never prayed before, only to Allah as a small child, and those had been recited prayers. I had no idea what to do next. I just knew it was time to surrender to what I knew in my heart was the truth: the man in white was Jesus. And he was worth following.

PART 4

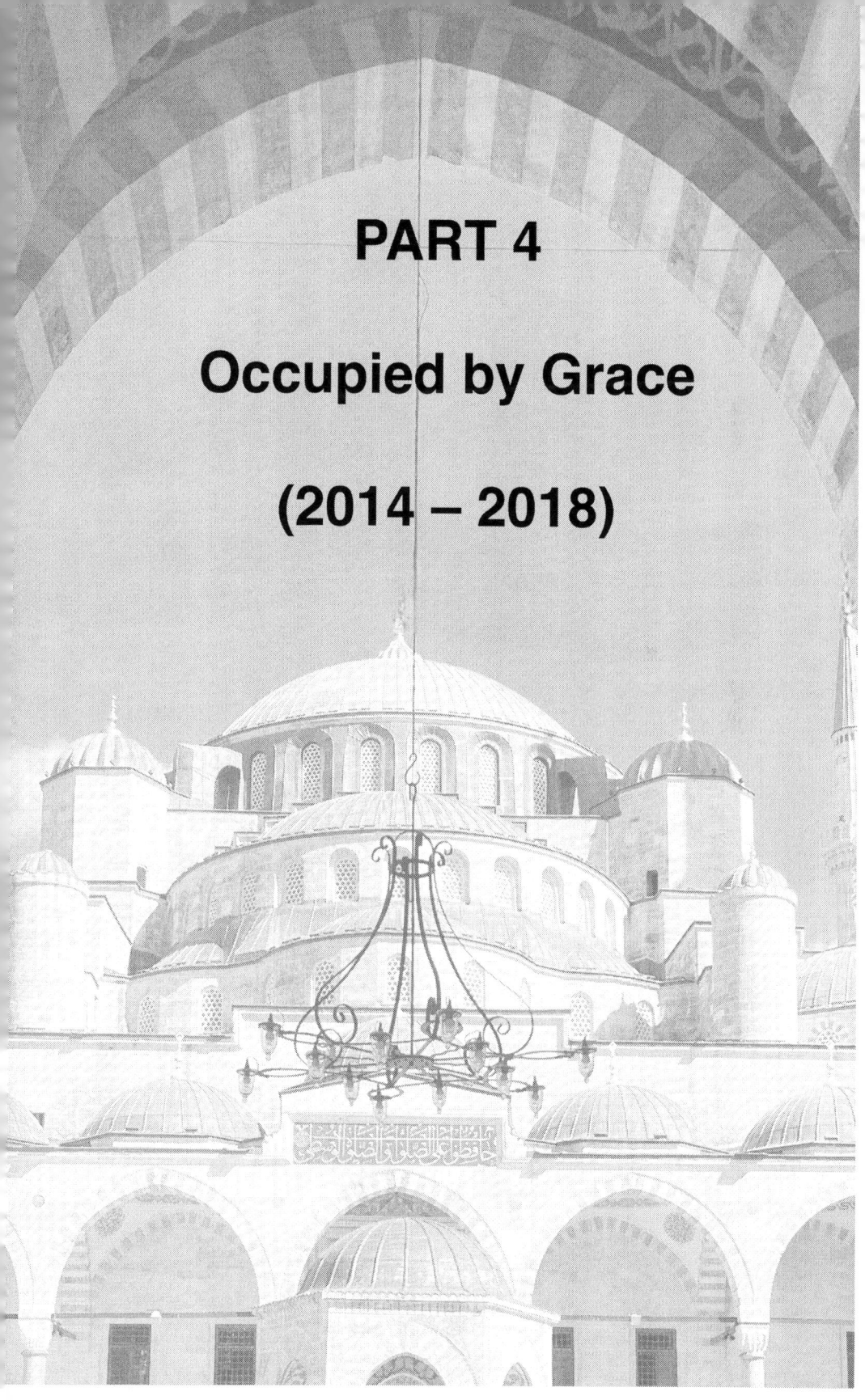

Occupied by Grace

(2014 – 2018)

21

LOVE COVERS ALL

As soon as I landed back in Raleigh, I went straight to Hannah's house. I couldn't wait to tell her what had happened in Turkey. She opened the door, surprised to see me standing on her doorstep. I smiled and didn't even pause for a hello.

"I met him!" I blurted out.

"You met who?" she asked.

"I met him!"

"Who?" she repeated, confused.

"I met Jesus." I said.

A smile broke across her face, and I saw tears well up in her eyes. She put her arms around me.

"Waleed, I'm so happy for you," she replied. "I can tell you met him. Your countenance has changed. I can see light in your eyes."

I did look different, and I also felt different. After ten years of living through war, peace was an unfamiliar feeling. And suddenly I had a deep sense of it. The wrestling in my heart was over.

That following Sunday, I attended church. And after the service, I ran up to tell Pastor Jim I had met Him. He hugged me and said it was the best decision I would ever make.

"What do I do now?" I asked him.

I still wasn't sure how the whole salvation thing worked.

He looked at me with a sideways smile.

"Waleed, Jesus has already done everything for you on the cross."

I remembered that golden cross in the Haggai Sophia and took a deep breath.

"That's it?" I thought.

In Islam, I always felt like I had to earn God's favor. Pray five times a day, fast during Ramadan. Was Christianity not the same? At first, I thought I might have to pray in a certain way, memorize the Bible, or fast to receive forgiveness and salvation.

I had never heard about the concept of grace until then.

"Waleed," Pastor Jim continued. "You can't earn anything. You will never be good enough. It's Jesus's sacrifice that bought your salvation. In ancient Greek, the word for salvation is 'sozo' which means deliverance, restoration, protection, preservation, healing, and being made whole.

"Jesus doesn't just save us so we can have eternal life and go to heaven; he saves us to live an abundant life, one where we are healed and walking in freedom on earth. One where we can fulfill the purposes he originally designed for us.

With my past, I thought I could never be good enough. The idea of grace—receiving something I didn't deserve—was completely foreign to me. I had been so arrogant before. But for the first time in my life, I saw my own sin and pride clearly. I had to turn away from my sin that separated me from God. I felt like the thief who hung next to Jesus on the cross. He was a murderer, a thief, a liar—just like me. Yet, Jesus had spoken a promise to him.

"Today you will be with me in Paradise."

Not because the thief was good enough, but because he'd surrendered and asked Jesus to remember him. That was grace.

The other concept that was hard for me to wrap my mind around was God being my Father and me being His son. I had never heard of God spoken of as a father who loved me and wanted me to be part of his family. God always seemed scary, far off, and unapproachable.

Pastor Jim gave me some reading material and talked to me about baptism. I was excited and also a bit terrified of what was next. Another layer was being peeled off. Now that I knew the truth, I started on a journey of developing my relationship with the God of the Universe, who was also my father, a relationship that was both thrilling and tender.

In those next weeks, I also learned more about the darkness I'd been so familiar with for so long. I now knew there was a real God; I also came to understand I had a real enemy who wanted to steal, kill, and destroy. Days after I had given my life to Jesus, the enemy of my soul—the accuser of the brethren—wanted to steal my peace and bombard me with fear and doubt.

I began to worry about what my family would think if I told them I was now a Christian. Would they disown me? Would they be safe if someone else in Iraq found out? Being an atheist had been a lot less risky.

I found myself in a lonely place. I didn't know any other Arab believers, so I had no idea where I belonged. I had rejected the religion of my birth, yet I still loved my culture and country. It was an unsettling place to be. I wanted to remain a part of my family in Iraq, but what about this new Christian church family? Despite all these thoughts swirling in my mind, I made a decision to follow Jesus no matter the cost. I also began to pray for my family to come to know the true God and to have a real encounter with Jesus like I did.

I began to get curious about baptism. Before, I thought baptism was just a physical process of getting dunked in some water. But as my faith grew, I began to understand it was not only a physical act but a spiritual act of transformation that washes away my old nature of guilt and shame and makes me a new creation. And I decided it was the next step for me.

I called Pastor Jim, and he organized it all. I invited all my friends who had been a part of my faith journey since moving to North Carolina. Hannah, Bill, Debi, and Marrow, my friend from language school, all came to witness this moment. As I stood in the small tub, water lapping at my legs, I couldn't help but wonder what would happen when I went under. Would I feel something miraculous? Would there be a sign or a sensation that marked the transformation?

Pastor Jim looked at me and smiled.

"I baptize you in the name of the Father, the Son, and the Holy Spirit."

When I was finally submerged, there wasn't any overwhelming revelation—just the feeling of water washing over me, soaking me to the bone. But as I was lifted out, something shifted inside me. It felt like new breath filled my lungs, fresh and pure. I realized that a new chapter had begun, and the weight of my past no longer held me down. I now had the choice to die to myself—my desires, my pride, and my past—in order to truly live. But this time living would be for Jesus, letting Him guide my path and define my purpose.

My conversion had nothing to do with Hannah, but God had used Hannah to point me to him. However, I'd be lying if I told you that deep down I wasn't holding on to a sliver of hope that she'd move me out of the friend zone—one day. But until then, I went back to Fort Bragg with my new faith in Jesus and new resolve to make it to the Q course and earn my green beret.

★ ★ ★

While on base, I started longing for a break from the busyness of the day so I could spend more time reading the Bible. So when I was assigned guard duty from 6 p.m. to 6 a.m., I was actually excited. I knew the nights would be quiet at the company HQ. I brought my Bible and one of Max Lucado's books with me and used the opportunity to find some peaceful time alone. The empty company hallways turned out to be the perfect place to read and reflect, free from distractions.

As I read the stories in the Bible, they started coming alive, and for the first time, I felt like I could truly hear God speaking to me through the scriptures. The Bible says, *The word of God is living and active.* Now, I was experiencing this.

It was incredible to hear God's voice as a gentle, still whisper.

"You belong," I heard Him say in one of those moments. It was the confirmation I needed, an assurance that I belonged to Him.

I loved what was happening to me—the conviction, the positive changes in my thoughts as I read these pages. My prayers now felt like a real conversation. It wasn't just me speaking into the void; I spoke to God, and He spoke back to me. All of it was transforming me into a new person. I was truly born again.

★ ★ ★

It was a windy day, and the conditions were not ideal for an airborne jump. But it had to be done; it was mandatory. As I exited the airplane with my T11 parachute, I caught a thermal pocket that lifted me higher, pushing me further away from the drop zone below. Desperately, I began pulling on my chute to escape the thermal pocket. By the time I broke free, I was swaying uncontrollably left and right.

As I neared the ground, I realized I was in a terrible position for landing. Instead of a proper landing, my buttocks absorbed the full blow of the impact. Then I rolled forward as the chute dragged me along with it. Pain shot through my body, and I couldn't move my legs. Raising my hand, I signaled for help. When the medics arrived, they took me straight to the hospital.

That night, I found myself in a hospital bed with level ten pain despite medication, diagnosed with a devastating back injury.

A few days later, despite my injuries, I was scheduled for a night jump. I asked to sit it out, but I had no choice. I jumped again, and upon landing, the pain in my back intensified. The next day, as I was getting out of the car, I collapsed. My legs were too weak to support me. I lost all feeling in them. I sat on the ground and drug myself towards the barracks where I leaned myself against the wall.

At that moment, I knew I would be forced to drop my airborne status, and with it, my dream of becoming a Green Beret slipped away. It was a heartbreaking realization that my body could no longer endure the demands of being a soldier. I felt broken and confused. If I wasn't a soldier, who was I?

★ ★ ★

During that time, Hannah and I began to spend more time together as friends. We often talked about God, our love of travel, and our work. One such day, we were at a coffee shop near her house. She was drinking hot green tea, and I was enjoying a latte, when she asked me a question that caught me off guard.

"Waleed, what is your dream, like your big dream for your life?" she asked.

I looked at her, puzzled.

"My dream?" I repeated and paused to consider. "Growing up, I never thought I'd live this long. My dream was just to survive."

I paused again, reflecting on how my latest dream of becoming a Green Beret had just crumbled. I leaned in, searching for the right words. But before I could, Hannah jumped in.

"Wow. I can't imagine growing up like that," she said. "I've had the privilege of having all sorts of dreams."

I didn't want Hannah to pity me, so I quickly continued. "I guess my dream now would be to be a good husband and father one day."

It was true—since all my family was back in Iraq, I longed for connection and the chance to build something meaningful here.

I noticed a shift in her body language, her expression softening. Since her mother passed away, Hannah had been adept at keeping people at arm's length, building a sturdy wall around her heart to avoid getting hurt. But something changed in that moment. As I shared my dream, I could almost see the first brick of her wall start to crumble.

A few days later Hannah called me, her voice a little nervous.

"Waleed, I know this sounds crazy, but I was listening to the song called *Worth It* by Francesca Battistelli, and basically, it talks about love. How it might be hard but it's worth it, and it can heal you. I felt the Holy Spirit fall on me, and I started crying. I didn't know what was happening, so I began to pray. And I felt a nudge to open up my heart. Before, I knew we could never be compatible, but now things are different."

I was freaking out. What was happening? Was Hannah saying she was willing to be my girlfriend? Before I could jump in, she continued.

"I want you to know, God has been slowly working on my heart. I know we come from two different cultures and have very different pasts—but really none of that matters if we are in Christ. And I know I have more work to do, I need to remove a few more bricks from my guarded heart."

I don't remember much more of the conversation except for finally blurting out my question.

"So are we dating?"

"We can go out on a date," she replied, emphasizing the 'a'. "We are taking it slow."

I was so ecstatic that I decided to drive straight up to Raleigh and take Hannah on our first real date that evening. We went to a Thai place around the corner from her house.

The dinner had a bit of an awkward vibe—I was trying to play it cool, and Hannah kept reminding me that we were just dating and not officially boyfriend and girlfriend yet.

"Baby steps," she said, looking at me seriously. "I still need God to remove the rest of the bricks."

At the end of the evening, I pulled up to Hannah's dad's house. She explained to me she'd been living with her Dad for the last two years while recovering from Lyme disease. She fidgeted with her dress, clearly embarrassed about being twenty-nine years old and living at her dad's house.

"I know it's not the most glamorous situation," she said, avoiding my gaze. "It feels like...high school."

I smiled, trying to ease her discomfort.

"In Iraq, it's completely normal to live with your parents even after you're married. There's no shame in it. In fact, can I come in and say hello to your dad?"

Hannah hesitated. I could tell she was calculating the risk.

"Yeah, I guess so."

At this point, Hannah didn't know that I had history with her father, and she just thought I was being polite. When we entered the house Hannah's dad was lying on the couch watching the nightly news as we slipped into the living room. We joined him and sat down on the coffee table beside the couch. I grabbed Hannah's hand.

"Pastor Jim, I just want you to know I really love your daughter, and I'm going to take good care of her."

Pastor Jim sat up intrigued.

Hannah elbowed me to stop but I continued with a bunch more nonsense before I ended it with a kicker.

"So, I was wondering if I could call you 'Dad?'"

Hannah flashed me a deadly look.

"What the crap," I thought. "What have I done."

Hannah had been telling me all night at dinner this was just a casual first date. We were not even exclusive.

Pastor Jim looked at me confused but smiling.

"Sure. I guess," he replied graciously.

Hannah's cheeks were red with embarrassment. She grabbed my hand, pulling me towards the door.

"Ok, it's time to call it a night!" she shouted.

She quickly rushed me out the door. Dazed, I sat on her front stoop for a minute. I could hear Hannah inside.

"Dad, what the heck," she raved. "He can't call you 'Dad!' People at church will think we are married. What were you thinking saying 'yes?'"

"Well," Pastor Jim replied. "I didn't know what was happening, and I didn't want to hurt his feelings. Maybe it's cultural."

Hannah shouted, "Cultural or not, he is not calling you 'Dad!'"

I felt doomed. What had I done? My dream relationship seemed over before it had even really begun. As I drove back to Fort Bragg, I couldn't stop thinking about whether I should call Hannah. Why had I said that? It didn't make any sense. I was sure Hannah thought I had taken things way too far, way too fast. How could I possibly recover from this?

The next morning, I was relieved when I saw Hannah's number pop up on my phone. Perhaps I could straighten this whole mess out.

"Waleed, I think we should talk. Are you free for lunch today?"

"Of course. I can be in Raleigh by noon."

At lunch, even before the meals arrived, before I could bring up the blundering mistake of last night, Hannah leaned in quietly.

"I know dating is not really a thing in Iraq, and you normally just get engaged," she began. "Is that why you asked to call my dad, 'Dad?'"

She was kindly trying to give me the benefit of the doubt by chalking it up to a cultural misunderstanding.

"No, Hannah," I replied. "Honestly, I've been thinking about it, too. I think I just got really nervous and didn't know what to say. And I word-vomited. I'm so sorry, I will never call him 'Dad' again."

I could tell she wasn't expecting that answer. I totally could have lied to her and said it was cultural—and taken the easy way out. But I think, the fact I told her the truth, that I had just been a bumbling idiot, made a few more bricks fall.

As we continued to go on dates, I found myself needing to disclose some very important parts of my life—beyond the obvious of having been married before. I needed her to understand my relationship with my family in Iraq.

One evening we were talking about our relationship deal-breakers, and I thought it's now or never. I looked at Hannah; she could tell I was getting ready to launch into something serious.

"Since my dad was killed in 2007, I have been the father figure to my siblings. I also support them and my mother financially every month. Is this an issue? Because I need you to know, I am committed to my family."

Hannah smiled.

"No, of course not," she replied. "It's actually refreshing to see a man take care of his family. And if I was in your position, I'd do the same. Family comes first."

Hannah then turned to me, looking nervous.

"I might have a deal breaker, too."

"Oh, no," I thought. I hoped it wasn't something really bad.

Hannah looked down to avoid eye contact. I could tell she was embarrassed.

"You know, I told you I have Lyme disease and have been really sick for over three years. What I didn't tell you yet was that I don't know if I'll be able to have kids. The infection went to my heart, and I had to have emergency surgery. That's when they found a seven-pound mass near my right ovary and fallopian tube. And they had to be removed. The doctors said I might not be able to get pregnant, or it could be very difficult for me."

Wow. That was not what I was expecting.

"Would not having kids be a deal breaker?" I wondered.

My mind raced to the baby I had lost; the pain was still there. Then I thought about Hannah, about how deeply I loved her, about how pure and genuine this love felt. I couldn't imagine my life without her.

I leaned over to hug her, pulling her close.

"Hannah," I said softly. "I love you. Kids would be great one day, but they're just the icing on the cake. You, my love, are the cake."

Over the course of the next year, God continued writing our love story, and it was more beautiful than I could have ever imagined. Hannah came into my ordinary life and turned it into a fairytale.

I was by now madly in love with her and determined to propose, but I didn't have any money. I dreamed of buying her a beautiful diamond ring. But with my meager army salary, and only a few months left on my contract before becoming jobless and homeless, it seemed impossible.

One day after church, Jenny asked how things were going. Her eyes twinkled with curiosity. It was clear she was wanting to know how serious things were between Hannah and me.

"Things are going really well, actually." I replied. "I'm planning to propose to Hannah." I admitted, unable to contain my excitement.

Jenny's eyes widened in delight.

"That's wonderful. But...how are you going to do it? Do you have a ring yet?"

I hesitated, then shook my head.

"That's the tricky part. I don't have the money for a ring right now. I'm trying to figure out a way to save up, but it'll take some time."

Jenny's smile grew even wider.

"You know," she continued, "I have a ring you could use. It was Mom's. Hannah always loved that opal ring. You can have it and use it to propose to her. Then you can buy her a diamond whenever you can afford it. That way, you don't have to wait."

I was taken aback.

"Are you serious?" I asked. "That's incredibly generous, Jenny. Thank you."

Here I was, wracking my brain for a solution, and suddenly, the Lord provided a way. When Jenny handed me the ring, I was overwhelmed. It was beautiful and held tremendous sentimental value. On one hand, I was glad to have it; on the other, I was terrified of losing it. It felt like I had a million bucks in my pocket. My excitement to propose intensified, and I wanted to do it quickly so I could give Hannah this ring she had always wanted. And so I wouldn't be responsible for it any longer than necessary.

It took me a few weeks to scout out the perfect spot for the proposal. I drove to different locations, visiting places that held special memories for us, like Duke Gardens and Historic Yates Mill Pond. Finally, I found the perfect spot overlooking the pond at Yates Mill. The scene was especially beautiful just before sunset.

I closed my eyes and envisioned the moment: I would walk her to this spot, and as she looked out at the view, I would drop to one knee behind her. When she turned around, she would see me there with her mother's ring in my hand, and then I would profess my love for her.

With the plan set, I could hardly contain my excitement. The ring felt like a precious secret, an amazing promise of our future together. And I couldn't wait to see the look on Hannah's face when I asked her to be my wife.

I nailed it.

The moment was all I'd hoped for. Hannah was totally surprised I had gotten her mother's ring. And she said 'yes!' With excitement she called her entire family, to share the good news.

Later that evening, she asked me if I would call and tell my family as well. I honestly didn't want to because I didn't know how they would react. They knew Hannah was a Christian, but they didn't yet know about my faith conversion. I was still trying to figure out how to tell them while weighing the risks involved. I felt like I was piling up a bunch of lies, and I hated being in that position. But I knew Hannah would not let this go. Family was super important to her and being accepted by her mother-in-law was high on her list.

So, I called, secretly hoping my mother would not pick up. But, she did. We shared our news and waited for her response. She wasn't thrilled about Hannah. And then we lost connection. I think she might have actually hung up on me. Hannah was devastated, especially because she didn't have her own mother anymore.

I tried to explain.

"She'll come around," I said. "It was just a lot for her. Put yourself in her shoes— imagine your son being far away in a totally different culture, getting married to someone you don't know with a completely different religion. How would you feel?"

With tears in her eyes, Hannah snapped back.

"I get it, but you married a Muslim already. And how did that work for you?"

We both started laughing.

"I promise she will come to love you, like I love you."

We got married a year and a half after our first date, exchanging our own vows in a meadow flanked with flowers on both sides. It was a perfect day, surrounded by beauty and love.

Just two months after our wedding, Hannah had the most vivid dream. In the dream, she was pregnant with a little girl named Grace. It felt like she would be the bridge that would somehow lead my family to the Lord. We thought maybe God would bless us with a child someday—

but we had no idea that only five months into our marriage, Hannah would actually become pregnant.

I remember the moment Hannah came down the stairs, holding the pregnancy test in her hand. She was trembling with a mix of excitement and fear. She didn't say a word but just held out the test to show me the two unmistakable lines. I stared at it, trying to process what I was seeing.

"What? Really?"

My voice was catching. Hannah just nodded, tears welling up in her eyes. Overwhelmed with emotions, she collapsed on the stairs, crying. I sat down next to her, unable to find the right words. We held each other, both of us crying. In that moment, we felt God poured His grace on us, answering our prayers and giving us a miracle we hadn't dared to hope for so soon.

22

THE GREAT RESCUE

My newly found dreams were coming true. I was a husband and now about to be a father. But life still felt bittersweet somehow. I missed my family and my brothers-in-arms. I often felt alone and insignificant as I transitioned into civilian life working as an IT analyst.

I often wondered, "Who am I, if I am not a soldier anymore?"

God was beginning to peel back another layer of my life.

Our daughter Grace was born under very stressful circumstances. Hannah's liver started to shut down at thirty-six weeks, so Grace was born via C-section a week later. By this point, I'd seen plenty of bloody carnage during the war, but something happened to me in that surgical room as I watched the doctors cut Hannah open and struggle to get Grace out.

On the outside, I kept my cool, but on the inside, I was freaking out. I saw my whole world lying on that table, and I felt like they were slowly slipping away. Finally, Grace emerged but was rushed away immediately—she was struggling to breathe. Hannah looked up at me helplessly as she was still being stitched up.

"I'll be okay," she whispered. "Go be with our baby."

I hesitantly left her to follow Grace as the nurse wheeled Hannah to recovery.

For the next six weeks, Grace had to be in the NICU—and it was a nightmare. She was born with neonatal pneumonia and then caught E. coli in the hospital. We were not allowed to stay overnight with her, so we'd spend most of the day there and stay late into the evening. Then we'd go home so Hannah could eat, pump, and rest for a few hours. By 3 a.m., Hannah was up pumping, and by 3:30 a.m., I'd be back at the hospital delivering the milk for Grace. Every night. Two weeks into this routine the entire night shift knew me as 'the milkman.' It was almost comical. And it was relentless.

I only had five days of paternity leave from my new civilian job, so a week into this grueling schedule, I had to go back to my day job. And it began to take a toll on my mental and emotional health. My nightmares and PTSD were intensifying, but I didn't want to tell Hannah. She had enough going on. We'd only been married for one year, and she'd been through so much, physically and emotionally. I needed to be strong. So I began to stuff down my emotions while silently unraveling inside.

We thought the NICU would be the hardest part of our new-parent journey, but as soon as we brought Grace home from the hospital, the wheels fell off. She wouldn't eat or sleep. She just screamed in pain pretty much all the time.

We went to a million specialists that all had differing opinions and all recommended intensive and expensive intervention protocols. After four months of running around to different medical appointments, while still not sleeping, of course, we were exhausted and confused.

One night Hannah and I were discussing the latest medical appointment. She explained this doctor had recommended to cut Grace's stomach open and do an exploratory surgery. Before she finished speaking, I jumped in.

"No. I'm not cutting my daughter's stomach open," I said, in a moment of clarity. "She doesn't need all these appointments, she just needs love and security. I think she has PTSD from her traumatic start. We're just going to keep her home and love her."

So, we decided to give all these medical appointments a break and to get on our knees. And we start interceding for Grace daily.

★ ★ ★

One month later, ISIS infiltrated Iraq, and my family was in grave danger. They desperately needed to flee from our home in Baghdad. All the borders surrounding Iraq were closed except for the border with Turkey.

My mother called, frantic, telling us they had less than twenty-four hours to cross into Turkey before its border would close, too. They needed $5,000 for travel expenses and visas. It was such a dire moment, and we didn't have $100 to our name, let alone $5,000. We were in a massive amount of debt from Hannah's medical bills, her school loans, and our wedding. In that moment, Hannah and I dropped to our knees, she looked at me, eyes full of determination.

"We may not have money," she said, "but we know the One who owns everything."

Together we cried out to God with all our hearts, seeking His guidance and provision. Praying for my family who meant so much to me. That's when the Holy Spirit reminded Hannah of something—a little thread that had almost been forgotten.

Just before her mother passed away, she had shared a dream of starting an organization called *Women Behind the Veil*, which would serve women and girls in the Middle East. At her mother's funeral, the family had asked for donations to *Women Behind the Veil*, instead of flowers. Now, it was Hannah's own mother-in-law and two sisters-in-law who were the women behind the veil and in grave danger.

Hannah called her dad.

"What ever happened to the *Women Behind The Veil* fund?" she inquired. "Did anyone donate?"

I could hear the hope in her voice. It had been a while since he'd looked, so her dad put her on hold to check the account status. When he returned, his amazement was audible.

"There is $5,000 in the account!"

We were in shock. Eight years after her death, Hannah's mother's desire to rescue the women behind the veil was becoming a reality. It felt like a miracle, and it was exactly what we needed to save my family.

I immediately wired the money to my mom, and then called her to tell her about the miracle and where she needed to collect the $5,000. It reminded me of Jesus's miracle of providing dinner for a whole crowd from five loaves of bread and two fish. It was supernatural provision.

My mom was crying with relief when she heard. We talked about the route they would take to Turkey and how they'd be able to reach the border. My military instincts were kicking in, and after we got off the phone, I started to feel anxious.

"I should be there to protect them," I thought. "They should already be in the US with me."

My mind was flooded with horrible what-ifs. I turned to Hannah; she knew that look. She grabbed my hand and squeezed it tightly to console me.

"Waleed, they're going to be okay," she said. "Let's pray for God to send them an angel to help them get across the border."

That had been my main worry, and that was exactly what we prayed for all night.

The next morning I was waiting for my mom to call when my phone finally rang.

"Yumma?"

"We've just arrived in Ankara, and we are registering with the UN as refugees."

In that moment peace flooded my heart.

Then she told me the craziest story. In an effort to avoid any ISIS presence on the ground, they had traveled by plane from Baghdad to Erbil in northern Iraq. Then they continued by bus to the Turkish border, arriving there around midnight. They were exhausted, scared, and didn't speak the language. My mom was seriously considering turning back, unsure of what to do next, when a man approached them, speaking in Arabic.

He asked her where they were trying to go, and she explained that they needed to get to Ankara, the capital, to register with the UN as refugees. The man nodded politely and led them to the ticket counter. Since he also spoke Turkish, he helped my mom, Fatima, Amel, and Mustafa purchase their bus fare. He even made sure they were settled on the bus before turning and walking away.

My mother looked out the bus window, hoping to catch his eye and thank him, but he had vanished. He was gone, just like that. I asked my mom what time that encounter happened, and I realized it was the very moment Hannah and I had begun praying for God to send an angel to them. It was a powerful reminder of God's faithfulness, showing me that, even when I couldn't be there to protect my family, He was watching over them and was taking care of them in ways I could never have imagined.

Another blessing materialized a couple days later. Hannah had traveled so much for work that she had saved up a ton of frequent flier miles. And I was able to get a free flight to Turkey, about a week later to see my family. I brought a suitcase full of winter clothes because, coming from Iraq, my family didn't have coats or sweaters. Hannah packed a very special red shawl to give to my mom, one that had been her mother's.

I was so excited to see my family. I hadn't seen my siblings in nearly seven years, ever since I had left in 2010. After touching down in Turkey, I grabbed my bags and rushed out of the airport, flagging down the first taxi I could find. I handed the driver an address I'd scribbled on a piece of paper, and off we went.

The city's streets blurred past as my anxiety mounted. Every second felt like an eternity, the anticipation of hugging my siblings and seeing them again overwhelmed me. I could hardly sit still. My mind was racing with memories and hopes, as I hurtled toward a reunion I had dreamed of for so long.

The taxi slowed down as we approached the apartment complex. I looked out the window and saw my little brother waiting on the sidewalk. My heart skipped a beat—he was no longer little; he was taller than me. As I got closer, I could see the years of trauma were etched into his face. A wave of sadness washed over me, knowing life had been unbearably hard for him. Guilt gnawed at me for leaving him all those years ago. But now, as I stepped out of the taxi, I was determined to make things right.

I immediately hugged my brother tightly, and neither of us wanted to let go. Finally, he grabbed my bags and led me to their new home, an apartment building where mostly refugees lived, both Iraqi and Syrian. The conditions were rough compared to American standards, but I was just grateful they weren't living in a refugee camp and that they were safe. The building was worn and crowded, but it was a place where my family had found some semblance of stability amidst the chaos of their lives. As we walked inside, I felt a mix of relief and resolve—I was home, at least for a few weeks.

My mother and sisters jumped up and hugged me as soon as I walked in. We were all crying. My heart finally felt like it was home. My sister made some chai, and we sat down in the living room to visit. I asked each of them how they had been. My sisters were a little distant and shy; they were young women now, and when I'd left, they had just been little girls. They didn't really know me anymore, and I certainly didn't know them as women. This revelation broke my heart. But I focused on the fact that I was here now and that I'd make it up to them.

That first night after I arrived, I FaceTimed Hannah so she could talk with my family. They were excited to see Grace, too. And to my surprise, my mom was wearing Hannah's mother's red shawl. Hannah started crying when she saw her, overwhelmed by the sight. It was incredible—

my mom was being covered by the eternal love Hannah's mom held in her heart for women behind the veil so many years ago.

Then Hannah had the chance to share the story of her mother with my mom, explaining her heart for women just like her and her daughters. This gesture marked the real beginning of their relationship, allowing my mom to see Hannah's heart and love for her beyond their different cultures, languages, and religions. It was a powerful moment, bridging the gap between their two worlds.

The next day, my first order of business was to take care of my family's essential needs so they could live comfortably in their apartment. I went shopping for mattresses, kitchen supplies, groceries, and other necessities. During this process, I began to notice the family dynamics had changed quite a bit since I'd last seen them. My father's death, my departure, and their status as war refugees had profoundly impacted them all. My sisters had taken on the roles of housekeepers, helping with cooking and cleaning. My mother's health was poor, so she relied heavily on the girls. My brother, being the only male in the house, had developed a sense of entitlement and took full advantage of his position.

The trauma had impacted each of us differently. And I had never seen it from their perspective until now. Fatima had to leave college and forgo her dream of becoming a psychologist. She was forced to become a mother figure to Amel. She was the glue that held everyone together, always trying to protect the family. The little girl I remembered was now a stoic woman, always scanning to figure out who was friend or foe.

My mother had become fragile, her heart physically weakened by the many years of grief. She constantly seemed to be in a state of fight or flight.

Not having a father figure had profoundly changed my brother. Once in Turkey, Mustafa became more interested in hanging out with his friends, playing pool, smoking cigarettes, and drinking alcohol. He dealt with the trauma by escaping.

My youngest sister Amel was fourteen when they arrived in Turkey. She had to drop out of school, which she loved. And here she couldn't yet speak the language and felt isolated. Amel desperately missed her late father and her home. And she began to struggle with severe depression and suicidal tendencies.

The evening before I left, Amel asked if she could talk with me privately.

"Of course," I said.

I could tell she was really nervous, which worried me.

"Do you remember the morning you left?" she asked. "You let me sleep in your bed with you."

I nodded, feeling a lump rising in my throat. She started tearing up.

"When I woke up, you were gone, and I was heartbroken. Why didn't you wake me up and say goodbye?"

I started to cry.

"I'm so sorry." I responded openly. "I don't know. At the time I thought it would be easier for you."

"I've always wondered why you left me."

All I could do was hug her and say I was so sorry. My heart felt shattered, and guilt consumed me. Knowing my mother and my siblings were now refugees, that their lives were in ruins to some degree because of my absence as well as the war, was unbearable. This was not what I wanted for my family, and I didn't know how to fix it. Buying them things wasn't going to change anything. I knew better than that. But I desperately wanted them to know that I loved them and that I was so sorry their lives were in such a state of limbo now.

★ ★ ★

Upon returning from Turkey, my PTSD ratcheted up to an unbearable level. Guilt consumed me for leaving my family again. The nightmares increased, and my sense of remorse magnified. I couldn't shake the feeling that I was responsible for the fact that my family had become refugees. They couldn't reunite with me in America because of the broken immigration system, and they couldn't live safely in their homeland because of the power vacuum created by the US invasion.

What had the war even been for? Life, in many ways, had been better under Saddam. I had never thought I'd admit that, but I was beginning to think democracy didn't work in every country. Yes, he had been a terrible dictator who ruled with an iron fist, but under his rule, we had safety, security, and infrastructure. Once the US occupation ended, Iraq was left with a huge power vacuum, which allowed groups like ISIS and other militant factions to terrorize and destabilize the country.

'Free Iraq' had been my mission as a naive college kid. After being part of the US war machine, on both sides, I realized war was not black and white. And it's often what happens in the gray areas that ends up defining you.

I'd given my life to Jesus but my struggle for freedom still continued. The enemy hurled constant attacks at me, riddling me with guilt, shame, depression, and unforgiveness. Nightly, I was reminded of the demons of war that still haunted me. I had once been a fearless, even dangerous, soldier who was involved in matters of life and death. Our missions had national security implications. Civilians would watch us on the news. Now, I was living *Groundhog Day*, doing mundane IT tasks on repeat at my job, working with self-centered people who were struggling with first-world problems. I was having an identity crisis. Who had I become?

The sound of a mortar round jolted me awake at 3:30 a.m. I sat up in bed, drenched in sweat. It was quiet in the room around me. Just like every other night, my recurring dream had gripped me. I had been caught in a firefight so vivid that it had felt real. And then, like every night, my weapon had jammed. And I woke up from the sound of a blast. My mind immediately went to the guys I had left behind in Iraq. What had happened to them? Were they fighting ISIS now? I wondered how many terrorists they had taken down.

A wave of nostalgia washed over me as I fantasized about being on a mission, rescuing hostages, and eliminating the bad guys. But then reality hit me like a cold splash of water. I had a wife now—and a daughter. I couldn't do war anymore. That part of my life was over.

Here I was, living the so-called American dream, which to me felt more like a nightmare. Trapped between the adrenaline-fueled past and the quiet, mundane present, I couldn't shake the feeling of being out of place. I was caught between two worlds—again.

I grabbed my phone and started to scroll Facebook to distract myself from these intrusive thoughts. I read a post: *Firas is dead. Killed by an ISIS sniper while fighting in Mosul.*

"What the hell?! Firas was a lion." I was sitting straight up in bed now.

He was one of the Delta Company squad leaders I'd done most of my missions with. I couldn't believe he was gone. The weight of guilt hit me hard, and my thoughts spiraled out of control.

"Did a new guy make a mistake and fail to cover him?" I wondered. "Or did Firas do something reckless?"

I couldn't wrap my mind around it. I didn't have any answers, just a growing sense of helplessness. And guilt—that I hadn't been there for him. I made my way to the living room, hoping that a change of scenery would help shake these dark thoughts.

But as I settled onto the couch, the intrusive thoughts took over, each one darker than the last, and a wave of sorrow and frustration threatened to drown me. I felt like I was suffocating, trapped in a darkness I couldn't escape.

"What have I become?" I asked myself, cradling my head in my hands. "What is happening to me? What am I doing here? I need to be there! I am stuck! I'm drowning! I can't take this anymore!!!"

The demons from the war had long been haunting me by night and taunting me by day. Now they began suggesting that there was a way out. They whispered that the only way to be free was to take my own life. This thought circled my mind like a vulture. I told myself that God would understand, that He would forgive me for wanting to stop the madness. Slowly, I began to plan out how and where I would do it, feeling a strange sense of calm as I mapped it out in my head.

The lyrics from the song "Fade to Black" flooded into my mind, each line feeling like a familiar friend urging me to give in, to fade into the darkness. Reality seemed distant, like a fleeting dream, and I couldn't find a single reason to hold on. I believed the lie I was hearing.

"They're all better off without you," the darkness whispered.

It was as though I was standing at the entrance of a tunnel, with darkness pressing in from all sides, closing in, suffocating me. The presence of something evil filled the space, the air thick with the weight of the demonic.

"Daddy!"

Grace's voice called out, snapping me out of my thoughts. She came over and sat next to me, her sweet presence a stark contrast to the heavy darkness that had been suffocating me. Then the first light of morning started to peek through the living room window, casting a soft glow across the room.

Grace settled on the floor by my feet, clutching her doll, and without any warning, she began to sing.

"You are my sunshine..."

Her little voice filled the room, so full of joy and innocence. Her exuberant singing sent an SOS flickering down that dark tunnel I'd been trapped in.

I was her sunshine. I made her happy. I was the one she counted on, and she loved me more than I could ever understand. She didn't want anyone to take her sunshine away. Her song was a lifeline, pulling me back from the edge.

Suddenly, I understood that no matter how I was feeling, I didn't want to devastate my wife, my family, and all my brothers who survived the war. Light was flooding back in and I could see. For the first time in what felt like forever, I saw a glimmer of hope. I broke down. I hugged my little girl and wept."God forgive me," I whispered. "Help me to get through these episodes of depression and deliver me from evil."

★ ★ ★

My story echoes a similar tale from two thousand years ago. The Jewish people, living under Roman occupation, were desperate for freedom, hoping for a political savior—a powerful king to deliver them. But Jesus didn't offer political freedom. Instead, He came and offered a greater, deeper freedom—a spiritual freedom that lasts forever.

I spent six years looking for freedom to come through the American occupation in Iraq. I fought for that freedom, sacrificing everything, only to end up trapped in a prison of my own making—one built by hatred, unforgiveness, guilt, and shame. It wasn't until I surrendered my life to Jesus that I took my first step toward true freedom. That surrender was just the beginning. What followed was a profound journey toward experiencing the fullness of salvation—*sozo*—which means, not only being saved but also being healed, delivered, and made whole.

Now, I have a new identity—I'm not a terp or a soldier, but a son, deeply loved by his Heavenly Father. I am no longer defined by the scars of battle nor am I occupied by pain. I am now occupied by God's grace.

EPILOGUE

Six years after my lowest moment, I found myself standing in front of a group of twenty-five veterans at a Mighty Oaks retreat, sharing my testimony. As I recounted the part about Grace singing "You Are My Sunshine," I looked out at the other veterans in the audience and noticed one man weeping.

Afterwards, he approached me, eyes still red as he pulled up his sleeve to reveal a tattoo. It was a tribute to that very song, a reminder that he was his family's sunshine and had fought to stay in their lives just as I had.

I had never wanted to share my story, because I had so much guilt and shame about so many things. I felt guilt over having killed people. Guilt for having survived when so many friends were killed. Guilt for having recruited so many friends as terps, only for them to be killed. Guilt for making a bad decision to marry, which then prevented me from bringing my mother and siblings to the US with me.

I still wrestle with questions about what is true, especially when it comes to my home country and my new country's involvement there. So many tragic years later, it's hard to be confident who was on the right side of history. I watch documentaries about the Iraq War and see senior US officers discussing risk analysis.

"The risks you took were with me, my family, my friends," I think to myself in those moments. "It was our lives, our jobs, our nation, and our home." And my heart aches.

But I am certain that, by working as terps, my friends and I saved countless civilian and soldiers' lives. By translating, we helped many people out of dangerous situations. Yet, at the same time, so many died. So many people I cared about—both American and Iraqi—lost their lives. It's hard to reconcile the loss and to accept that things aren't black and white. Once again, I find myself navigating the gray area, but this time I do so in the light of Christ.

The enemy of my soul no longer has a hold over me. I have found freedom in Christ and now live with a new identity—not as a terp nor as a soldier, but as a beloved son, loved by his Heavenly Father. Becoming a Christian didn't instantly heal me from PTSD or rid me of the suicidal ideation that was driven by guilt and shame. It has been a ten-year journey of deliverance and healing through which I've discovered true 'sozo' life. I no longer live in an occupied nation, and my heart is no longer occupied by pain. Instead, it is filled with God's grace.

This past year, many years after my military service, my wife and I felt a clear call to step into ministry and missions. So, after much prayer, we made the decision to leave our comfortable jobs, pack our bags, and enroll in a Discipleship Training School (DTS) with Youth With A Mission (YWAM).

For five months, we embarked on a journey that would forever change the course of our lives—spending the first half of the time in a classroom setting and the second half serving in South Africa. During this time, my faith grew in ways I could never have imagined. I had the privilege of sharing the Gospel with others and witnessing God move in miraculous ways. I also experienced my own deliverance and healing from the wounds of my past.

Leaving behind my career and stepping into full surrender wasn't easy, but it birthed a new purpose in my life: to see other veterans set free—free from the chains of their past, just as God had set me free.

I now have the incredible honor of helping other veterans experience true freedom through the ministry I founded called Redefined Warriors. Our mission is to guide veterans toward discovering their true identity in Christ and embracing God's purpose for their lives. The saying 'once a soldier, always a soldier' might be true, but once you give your life to Christ, your life's mission shifts to something far greater than you could ever imagine. At least that was my experience. It's a higher calling, one that brings freedom, healing, and purpose beyond the battlefield. If you are struggling and need hope don't fight alone, please reach out to me. You'll find my contact information at redefinedwarriors.com.

My extended family is still living as refugees, caught in limbo as their paperwork has not progressed in a decade. We continue to fill out mounds of paperwork and to petition US officials. But so far, nothing has come of it. We know that God is bigger than the US immigration system, and we know He can do anything. So we continue to pray for the day when we can all be reunited, holding on to hope that God's timing is perfect. We have also had the privilege of sharing the Gospel with my family, and we look forward to the day when they surrender their lives to Jesus and reorient to true north.

Hannah continues to write books and create films that amplify light and inspire hope. Together, we are raising our two beautiful children to be Kingdom citizens—traveling the world with us on a mission to spread the Gospel of Hope to all people.

Our prayer for you, dear reader, is that you, too, may find God's grace taking up the space in your heart that has been occupied by pain. May His love guide you to true north, bringing you hope, freedom, healing, and purpose.

Salvation Prayer

Heavenly Father,

I come before You as a broken person. I have stumbled every step of the way, but today I surrender my life to You. I repent of my past sins and offer my present and my future to You. I believe that Jesus Christ is Your Son, that He died for me, rose from the dead, and is alive today. I invite You, Jesus, into my heart and life. Take over completely and root my identity in You alone. Shape and mold me into the person You created me to be. Help me to trust and follow You all the days of my life.

In Jesus' name, Amen.

My Prayer for You

Heavenly Father,

I pray that You would seal the hearts of those who have chosen to follow You. Surround them with Your hedge of protection, and do not allow the enemy to steal, kill, or destroy the seed of salvation that You have planted in their lives. Fill them with Your Holy Spirit, and let them walk fully in their salvation, experiencing Your healing and deliverance.

Jesus, guide them as they journey on this new path. Raise up mighty warriors who will stand firm, dismantling the enemy's influence in their lives and in the lives of their families. Help them reclaim their God-given identity and purpose.

In Jesus' name, Amen.

Camp Justice, Baghdad 6/1/2004
First helicopter mission
(Chapter 8)
Photo by Waleed Hamza

BIAP Area IV 3/26/2006
Just before the ICTF QRF mission at Sadr City
(Chapter 10)
Photo by Waleed Hamza

Shu'la District, Baghdad 4/21/2008
Humvee blown up by an EFP
(Chapter 13)
Photos by Wil Ravelo

"Cross Sabers" Park, Baghdad 3/20/2008
Hanging out with Wil waiting for go time
(Chapter 13)
Photo by Wil Ravelo

Muqur, Afghanistan 5/22/2012
My deployment with 82nd 1-504 ABN Div B.co
(Chapter 18)
Photos by Waleed Hamza

Post Deployment Ball, Fayetteville, NC 10/12/2013
I was still deep in the friend zone
(Chapter 20)
Photo by Mike MacLeod

Starlight Meadows Venue, Burlington, NC 4/6/2014
Our Wedding Day
(Chapter 21)
Photo by Rebecca Cain

Blue Mosque, Istanbul Turkey 7/22/2023
Standing in the same spot where I first accepted Jesus
(Chapter 20)
Photo by Hannah Hamza

YWAM Kona, Hawaii 9/1/2023
Attending Crossroads DTS
(Epilogue)
Photo by Hannah Hamza

Acknowledgments

We want to express our deep gratitude to the people who generously took the time to review this book—Michelle Yancey, Wil Ravelo, Jordan Jackson, Pete Kofod, and N.R. Your insights and feedback have been invaluable in shaping this story.

To Carlene, our editor—thank you for always pushing us—even when it was hard—and for your incredible work editing this book. You've been a true cheerleader during the toughest times, and we couldn't have done this without your support.

To Antje, our editor—thank you for guiding us to the finish line. Your meticulous attention to detail and sharp eye for every typo and missing comma continually amazed us.

To Zan, our designer—your creative genius and dedication to perfection brought the cover to life. We're so grateful for your passion and the care you took with every detail.

Thank you Bill and Debi Micheals for opening your home and making me part of your family. I'm grateful for your love and for showing me Jesus through your actions.

Thank you, Pete Kofod, for helping me navigate my new life in America and for being my mentor.

To Kris Johnson, thank you for sponsoring my visa and working tirelessly to bring me to America. Your unwavering support and dedication means the world to me. I am deeply grateful that you didn't leave me behind.

Thank you, David Nasser, for sharing your story of redemption, which opened the door for me to meet Jesus. It's by the blood of the Lamb and the word of our testimony that we overcome.

And to our extended family, thank you for babysitting our kids for countless hours so we could write and for encouraging us to tell our God story.

Lastly, a special thank you to my brothers-in-arms, who shared their stories, bravely revisiting some of the most difficult experiences of their lives. Your willingness to open up, even at the risk of triggering painful memories, has made this book possible. This is as much your story as it is mine.

Glossary of Acronyms

AAR – After-Action Review

AIT – Advanced Individual Training

ANA – Afghan National Army

AO – Area of Operations

BIAP – Baghdad International Airport

CCP – Casualty Collection Point

CIB – Combat Infantry Badge

HHC – Headquarters & Company Headquarters

COL – Military Rank of Colonel

COP – Combat Outpost

CPT – Military Rank of Captain

FOB – Forward Operating Base

HQ – Headquarters

ISOF – Iraqi Special Operations Forces, the Iraqi equivalent of US Green Berets

ISR – Intelligence, Surveillance and Reconnaissance

LT – Military Rank of Lieutenant

LZ – Landing Zone

MAJ – Military Rank of Major

NCOIC – Non-Commissioned Officer In Charge

ODA – Operation Detachment Alpha, an elite team of highly trained soldiers

OEF – Operation Enduring Freedom, US military action in Afghanistan from October 7, 2001 to August 30, 2021

OIF – Operation Iraqi Freedom, the US military action in Iraq from March 20, 2003 to December 15, 2011

POC – Point of Contact, the US military member to whom a terp reports

RASP – Ranger Assessment and Selection Program

ROEs – Rules Of Engagement

RTB – Return To Base

RTO – Radio Tactical Officer

SF – Special Forces, US Special Forces or Green Berets

SFC – Military Rank of Sergeant First Class

SGT – Military Rank of Sergeant

SITREP – Situation Report

SSE – Sensitive Site Exploitation

TOC – Tactical Operations Center

TOT – Time On Target, the anticipated arrival time at a target location

VBED – Vehicle-Borne Explosive Device

XO – Executive Officer

About the Authors

Waleed Hamza is a father, husband, and outdoor enthusiast. He served as a translator for the US Special Forces in Iraq and later, as a soldier, with the 82nd Airborne Division in Afghanistan. After his military service, he embarked on a profound journey of faith, which led him to establish **Redefined Warriors**—a ministry focused on helping veterans rediscover their purpose, identity, and dignity.

Hannah Sink Hamza's faith and creativity are the through-line of her life and career, putting her at the intersection of storytelling, motherhood, and missions. She holds the prestigious titles of mom and wife; not to mention her professional titles, award-winning filmmaker, published author, and creative director. Hannah has spent the last 20 years studying the power of storytelling and developing narratives that amplify light and inspire hope.

Connect with us.

waleed.hamza80
hannahhamzacreative

Made in the USA
Middletown, DE
31 October 2024